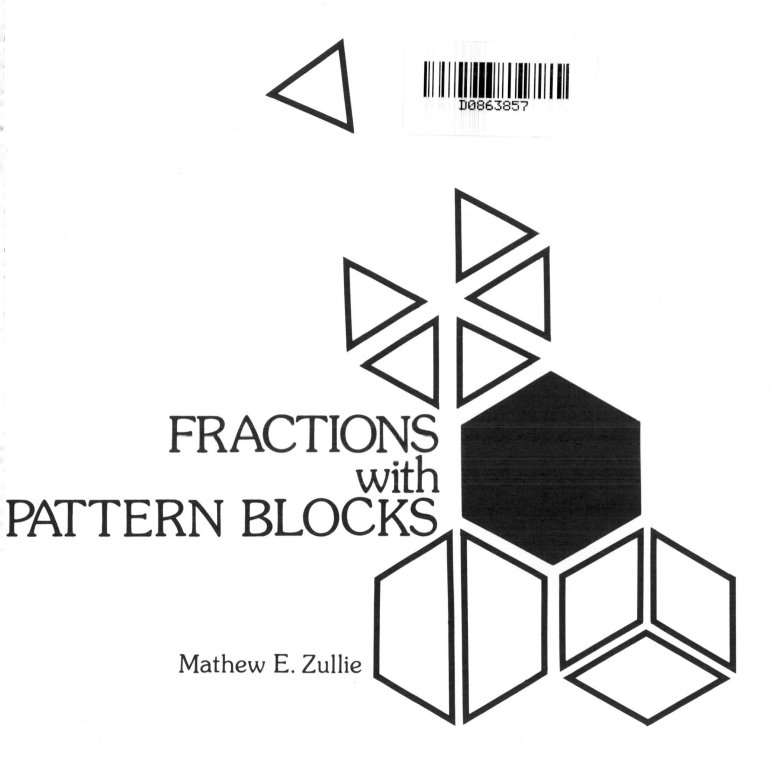

FRACTIONS
with
PATTERN BLOCKS

Mathew E. Zullie

Creative Publications, Inc.   Palo Alto, CA.

© 1975, Creative Publications, Inc.
P. O. Box 10328
Palo Alto, California 94303

ISBN:0-88488-050-8

12 . 8987

INTRODUCTION

This book of activities is intended to help teachers and students understand concepts concerning fraction relationships and operations with fractions. The book includes a series of activities to be used with a set of wooden blocks called Pattern Blocks. Pattern Blocks were developed by the Elementary Science Study group at the Education Development Center, Inc. in Newton, Massachusetts and are now available from Creative Publications and other mathematics materials companies.

This book presents pattern blocks as one of the many models for understanding the concept of fraction. The activities in the book develop the model to help the student visualize and concretely manipulate what is happening in the numerical operations. Emphasis is on development of the concept of fraction and of operations with fractions rather than on the transition from manipulatives to the written algorithms. In essence, the book presents the operations with fractions in concrete form so that the student may incorporate the concepts in terms of the manipulative. This depth of understanding may serve as a useful reference when questions later arise.

Fractions with Pattern Blocks is divided into two parts. The first part provides more than 50 student worksheets relating to the understanding of the concepts of fraction and fraction relationships. The second part deals more directly with the fundamental operations of addition, subtraction, multiplication, and division of fractions.

Pages are perforated so that they can be easily removed from the book and used as duplicating masters or placed in activity card protectors and used in a math lab setting. Please see the copyright page for the reproduction permission which has been granted.

In several activities, recording answers involves tracing blocks or using a pattern block template. Since it is sometimes awkward to hold the block and trace it templates are often convenient. These are available through Creative Publications. Most activities are appropriate for students with a wide range of abilities. It is important for all students to feel comfortable and familiar with the blocks before using them in the more advanced directed activities.

Most students enjoy working with the pattern blocks. Teachers have found them to be a successful teaching tool in remedial instruction, in introducing new concepts, and in drill and practice. You may wish to create activity sheets of your own to extend popular or especially fruitful lessons.

TABLE OF CONTENTS

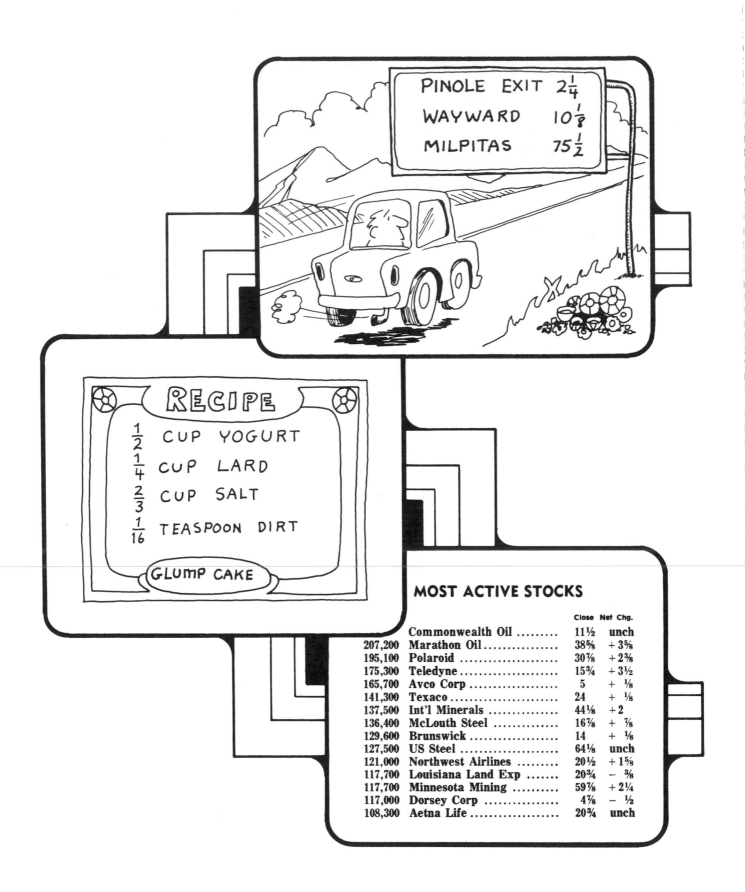

PINOLE EXIT 2¼
WAYWARD 10⅛
MILPITAS 75½

RECIPE

½ CUP YOGURT
2¼ CUP LARD
⅔ CUP SALT
1/16 TEASPOON DIRT

GLUMP CAKE

MOST ACTIVE STOCKS

		Close	Net Chg.
	Commonwealth Oil	11½	unch
207,200	Marathon Oil	38⅝	+ 3⅝
195,100	Polaroid	30⅞	+ 2⅜
175,300	Teledyne	15¾	+ 3½
165,700	Avco Corp	5	+ ⅛
141,300	Texaco	24	+ ⅛
137,500	Int'l Minerals	44⅛	+ 2
136,400	McLouth Steel	16⅞	+ ⅞
129,600	Brunswick	14	+ ⅛
127,500	US Steel	64⅛	unch
121,000	Northwest Airlines	20½	+ 1⅝
117,700	Louisiana Land Exp	20¾	− ⅜
117,700	Minnesota Mining	59⅞	+ 2¼
117,000	Dorsey Corp	4⅞	− ½
108,300	Aetna Life	20¾	unch

2

YOU MIGHT BUILD:

A TOWER or A DESIGN or A ROCKET

or A ROAD or A PERSON or ANYTHING

3

Cover with pattern blocks.

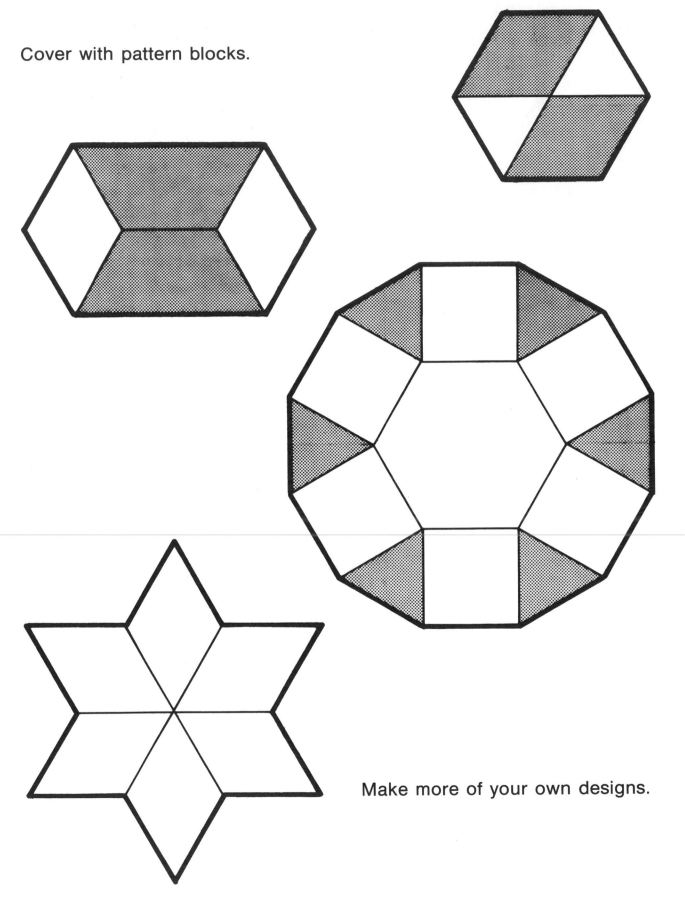

Make more of your own designs.

4

All of the activities in this book use pattern blocks to represent fractions. Their names are:

GREEN
TRIANGLE

BLUE
PARALLELOGRAM

RED
TRAPEZOID

YELLOW
HEXAGON

TAN
PARALLELOGRAM

ORANGE
SQUARE

Take out a set of pattern blocks before you continue.

Pattern blocks

How many △ green triangle cover ⬡ red trapezoid ? _____

How many △ cover ◇ blue parallelogram ? _____

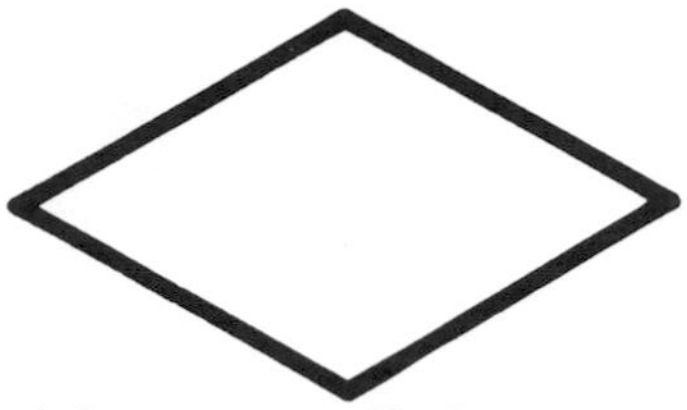

How many ◇ cover ⬡ yellow hexagon ? _____

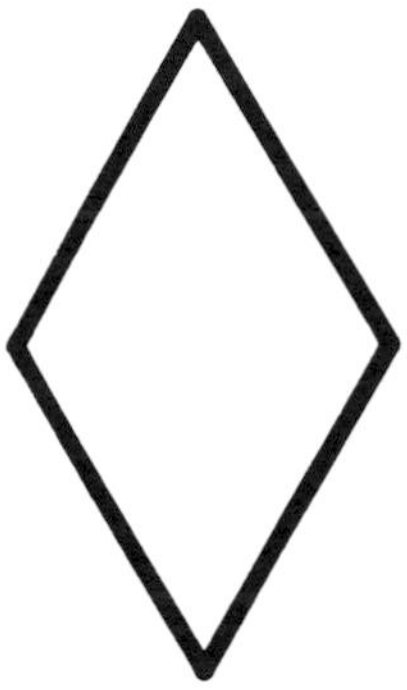

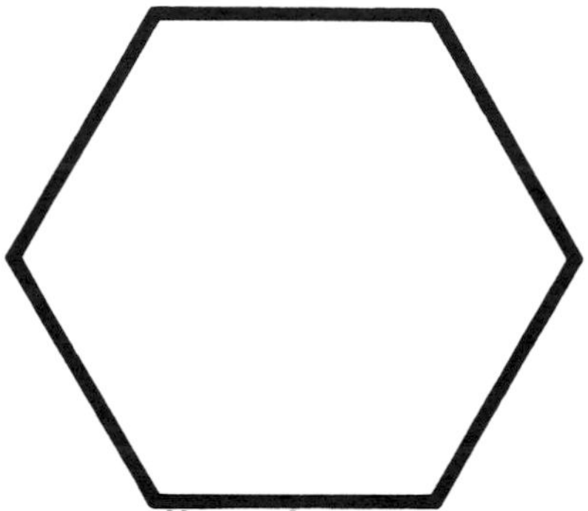

How many ⬡ cover ⬡ cover ? _____

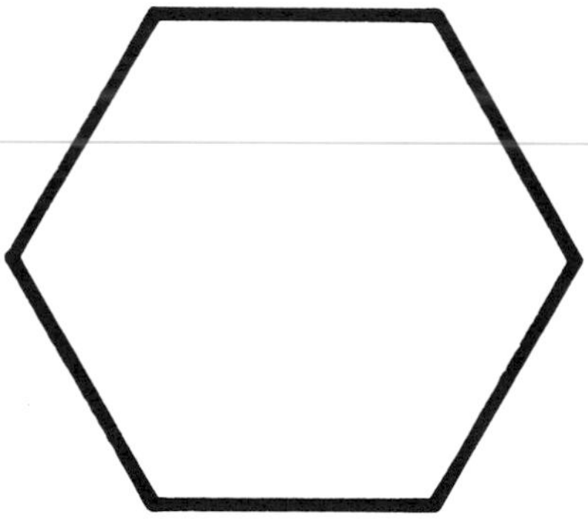

How many △ cover ⬡ ? _____

6

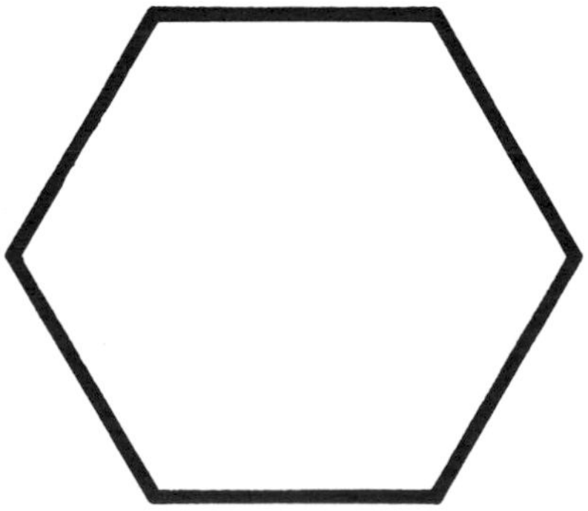

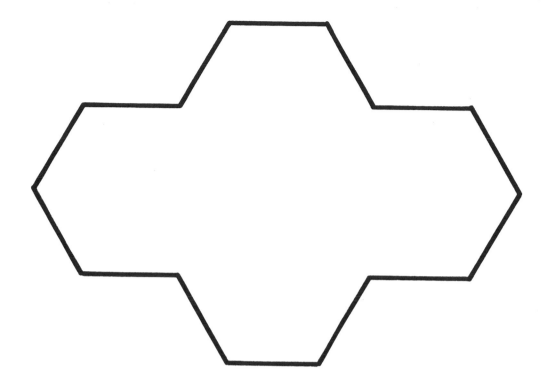

Cover the design with yellow hexagons (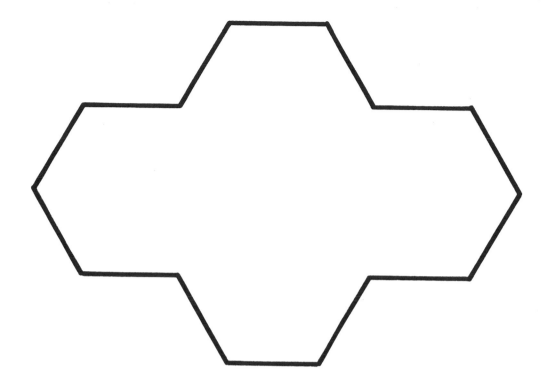).
How many hexagons cover it? _____

Cover the design with red trapezoids ( ).
How many trapezoids cover it? _____

Cover the design with blue parallelograms ().
How many parallelograms cover it? _____

Cover the design with green triangles ( ).
How many triangles cover it? _____

7

A *fraction* describes a part of a whole.

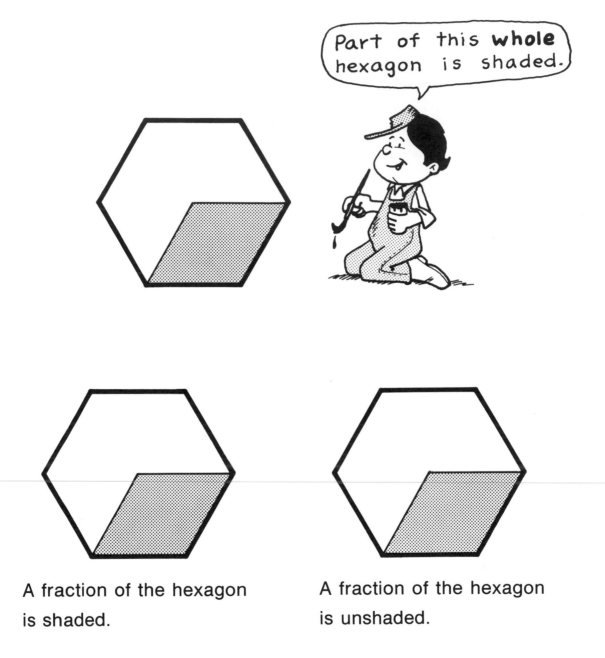

Part of this **whole** hexagon is shaded.

A fraction of the hexagon
is shaded.

A fraction of the hexagon
is unshaded.

Cover a fraction of a yellow hexagon with one blue parallelogram as
in the drawings above.

Now use other pattern blocks to cover a fraction of a yellow hexagon.

8

Cover the shaded fraction of each figure with green triangles.

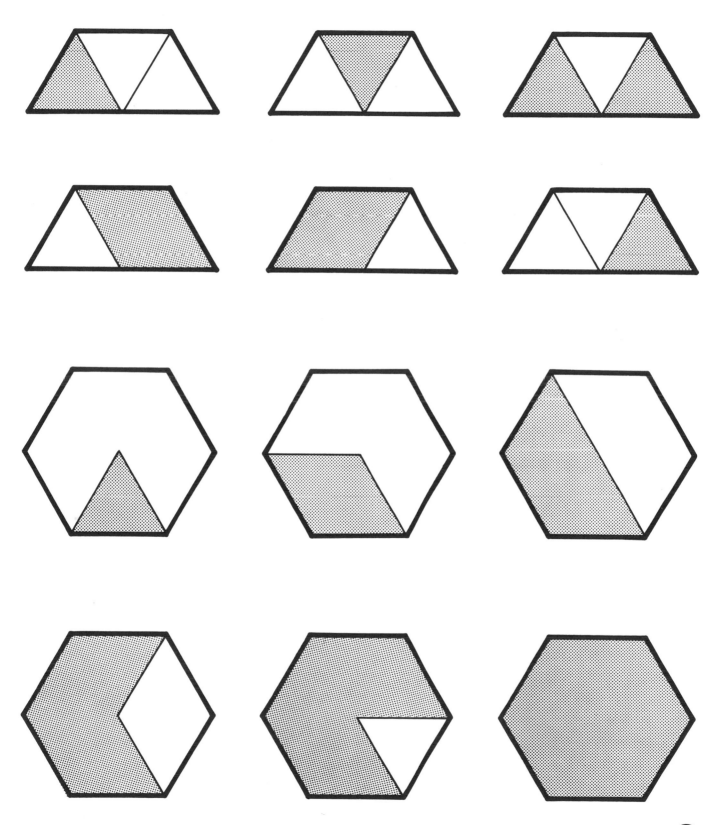

9

Cover the shaded fraction of each figure with blue parallelograms.

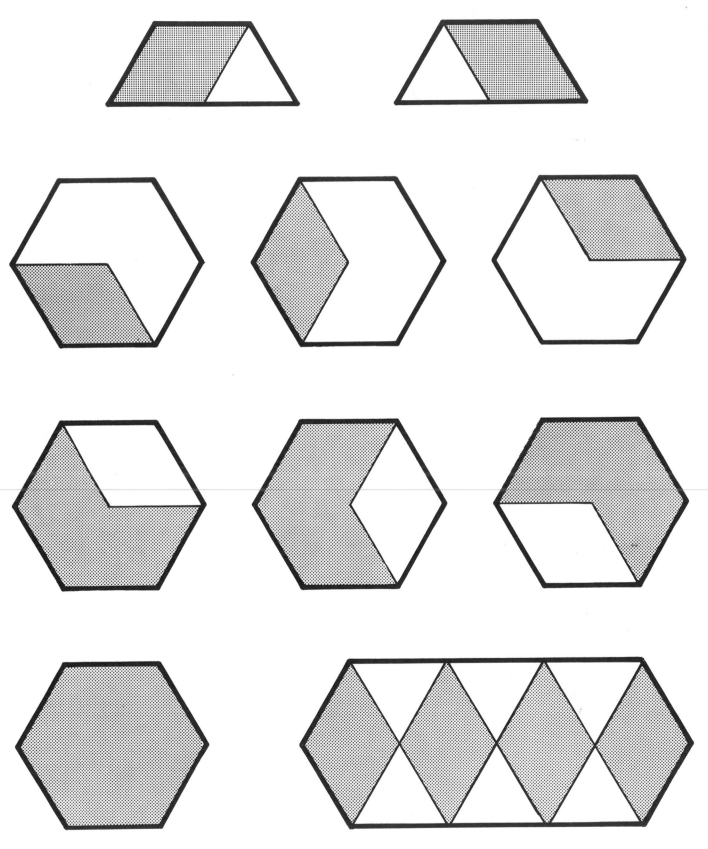

10

Cover the shaded fraction of each figure with red trapezoids.

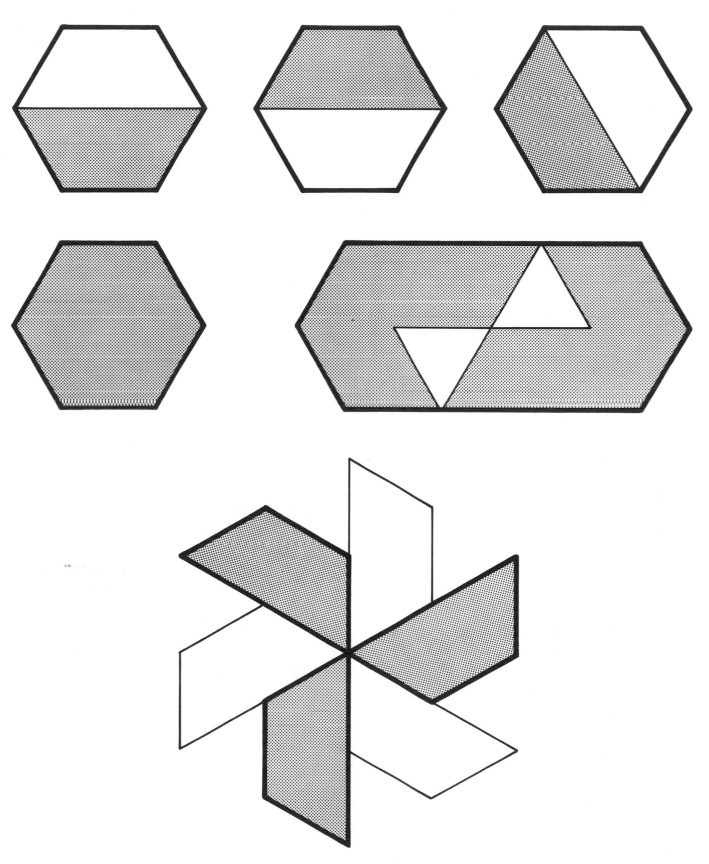

Cover the shaded fraction of each figure with orange squares.

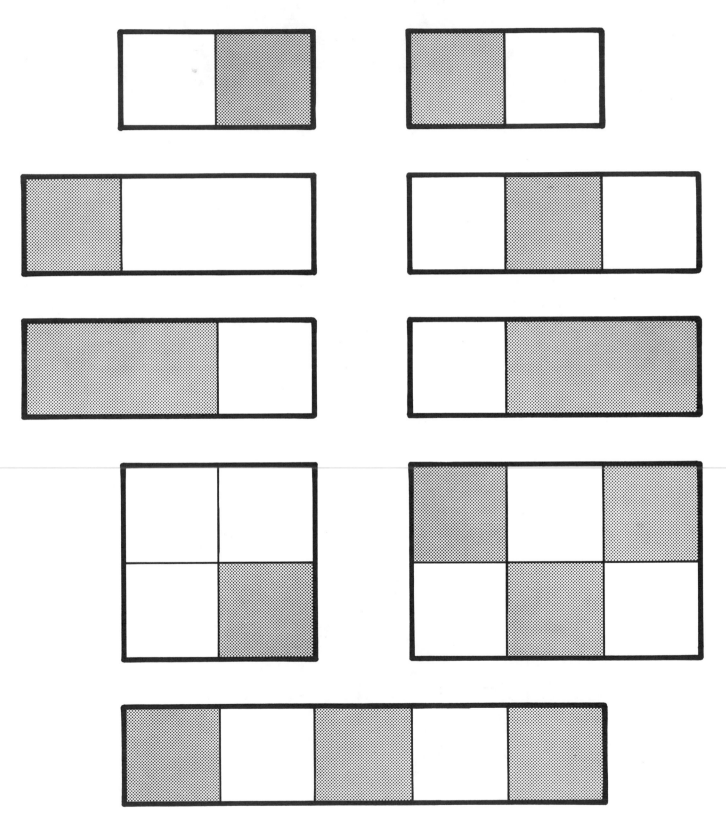

Cover with 2 blocks

Cover with 1 block

Cover with 3 blocks

Cover with 6 blocks

You may need to use different colored pattern blocks below.

Can you do this **2 different** ways?

It's ruff.

Cover with 4 blocks

Cover with 5 blocks

BLAST OFF

Number of Players: 2–4

Materials: 1 die; collection of blue and yellow pattern blocks;
   1 hexagon tree playing board for each player.

Rules: Players take turns rolling the die and drawing as many blue
   blocks as the number rolled. If possible, the player trades blue
   blocks for yellow. Yellow blocks are then placed on the
   hexagon tree. The first player to fill his playing board with
   yellow blocks is the winner.

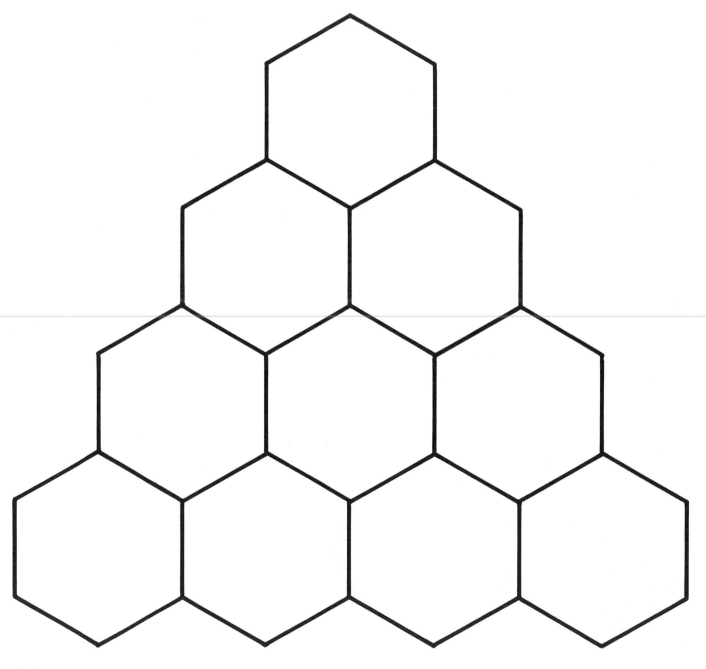

14

Each figure below is separated into two equal parts. Each part is (1) of the two (2) equal parts. We call it one-half and write it $\frac{1}{2}$.

Cover one-half $\left(\frac{1}{2}\right)$ of each figure with pattern blocks.

Write the numeral for one-half on the half you didn't cover.

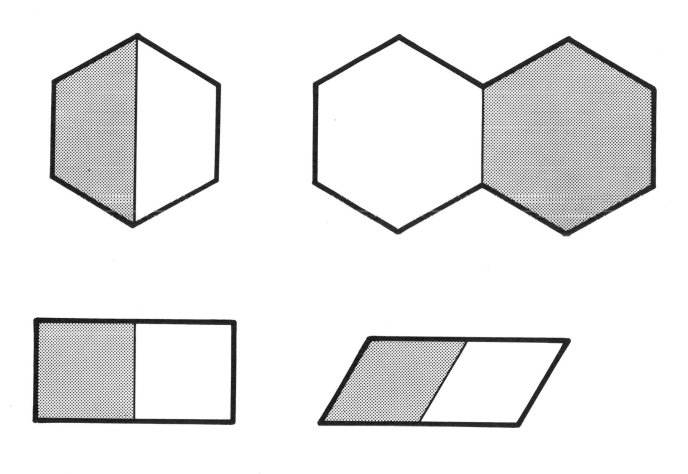

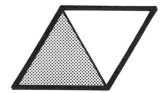

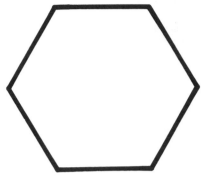

Cover $\frac{1}{2}$ red

and $\frac{1}{2}$ green.

Cover one-half red
and one-half green.

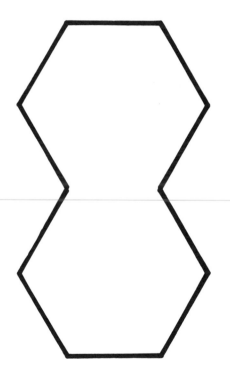

Cover $\frac{1}{2}$ red

and $\frac{1}{2}$ green.

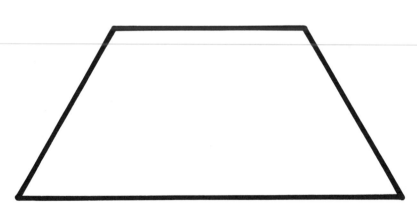

Cover $\frac{1}{2}$ red

and $\frac{1}{2}$ green.

Do you think the green blocks and the red
blocks in this figure cover the same
amount? _____ Why?

16

Cover one-half of each figure with blue and one-half with green.

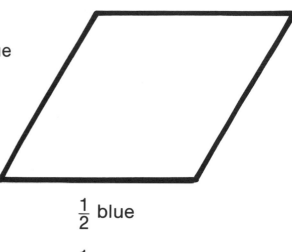

$\frac{1}{2}$ blue

$\frac{1}{2}$ green

After solving each puzzle, cover the blue blocks with the green blocks to see if they are the same amount.

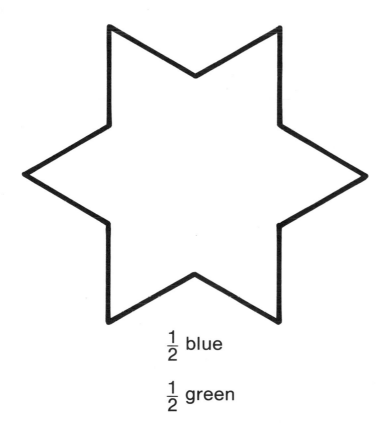

$\frac{1}{2}$ blue

$\frac{1}{2}$ green

$\frac{1}{2}$ blue

$\frac{1}{2}$ green

17

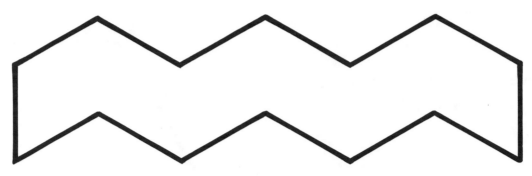

Cover $\frac{1}{2}$ green and $\frac{1}{2}$ blue.

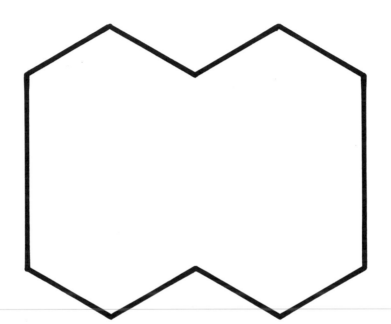

Cover $\frac{1}{2}$ blue and $\frac{1}{2}$ green.

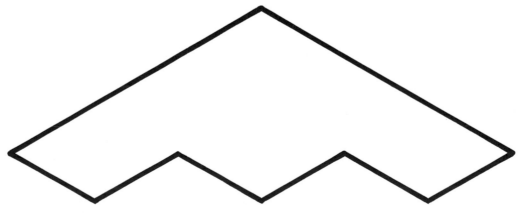

Cover $\frac{1}{2}$ green and $\frac{1}{2}$ blue.

18

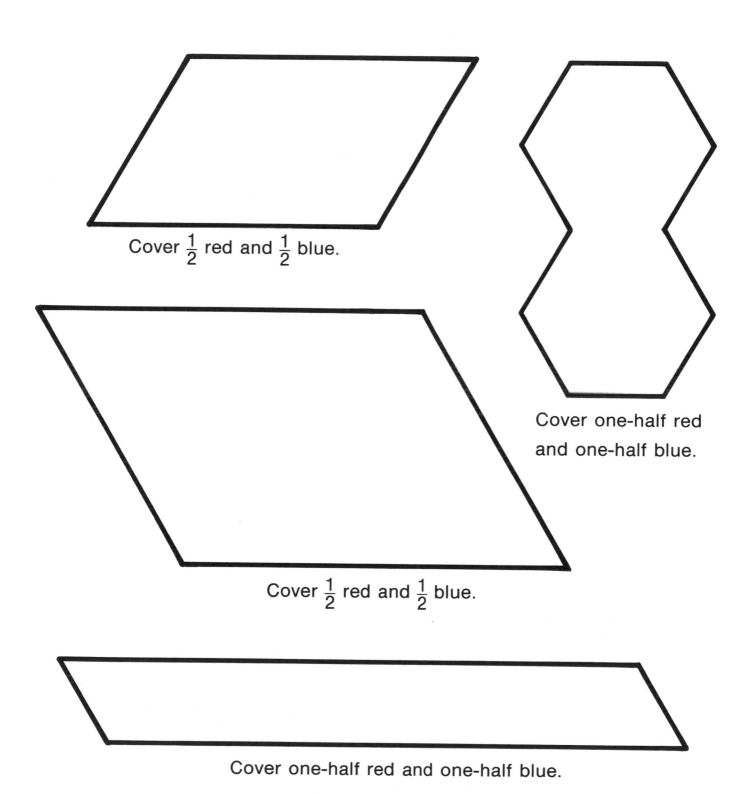

Cover $\frac{1}{2}$ red and $\frac{1}{2}$ blue.

Cover one-half red
and one-half blue.

Cover $\frac{1}{2}$ red and $\frac{1}{2}$ blue.

Cover one-half red and one-half blue.

19

Cover with 3 blue pattern blocks.

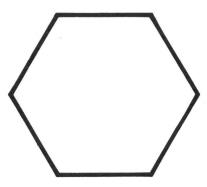

Each block is 1 of 3 equal parts or $\frac{1}{3}$ of the figure.

Shade $\frac{1}{3}$ of the figure below.

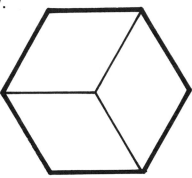

One of 3 equal parts in this figure is shaded. How many parts are not

shaded? _____ The unshaded parts are _____ of _____

equal parts of the figure, or $\frac{2}{3}$ of the figure.

Cover $\frac{1}{3}$ green and $\frac{2}{3}$ blue:

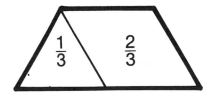

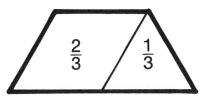

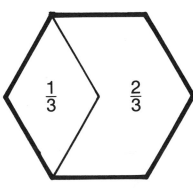

20

1) Cover each figure with 3 blocks of the same color.

2) Shade $\frac{1}{3}$ of each figure.

3) Cover $\frac{2}{3}$ of each figure with yellow blocks.

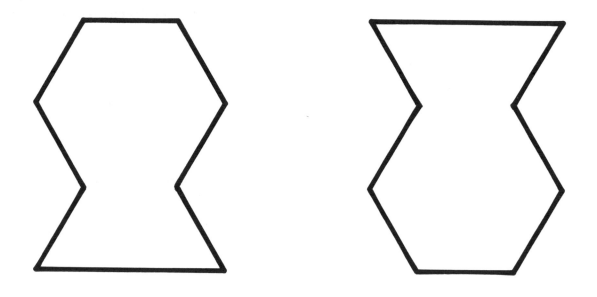

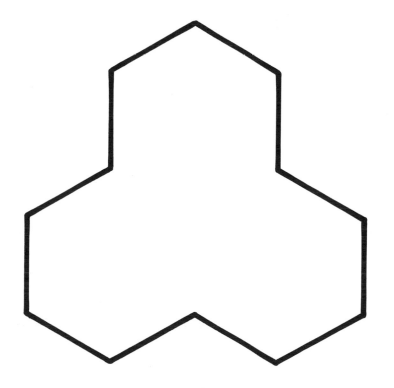

21

Cover $\frac{1}{3}$ of each figure with blocks.

Next, cover $\frac{2}{3}$ of each figure or _____ of _____ equal parts.

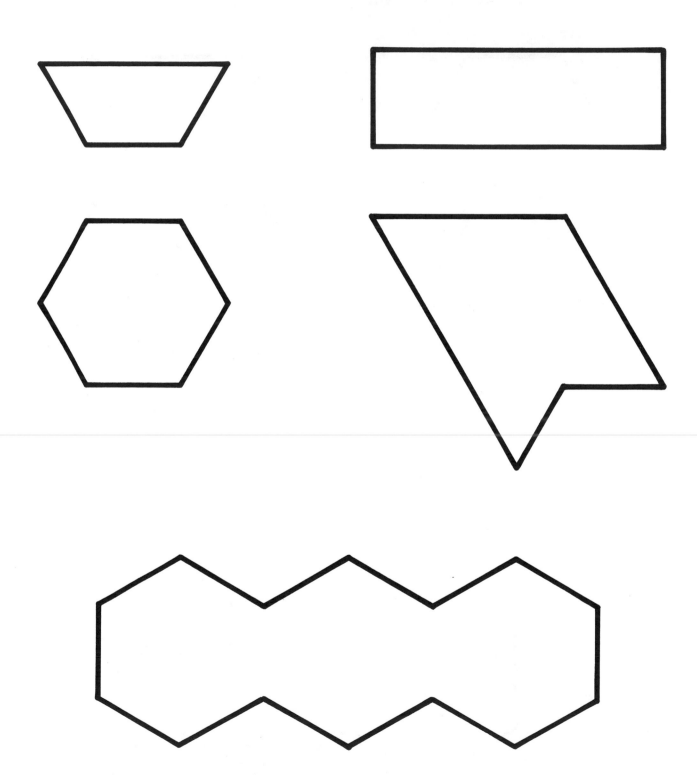

WHAT THE HEX

Number of Players: 2

Materials: Red trapezoids, blue parallelograms, and green triangles.

Rules: Each player in turn places one pattern block inside the outline below.

No pattern block may be placed across a solid line.

The player who places the last shape in the outline is the loser.

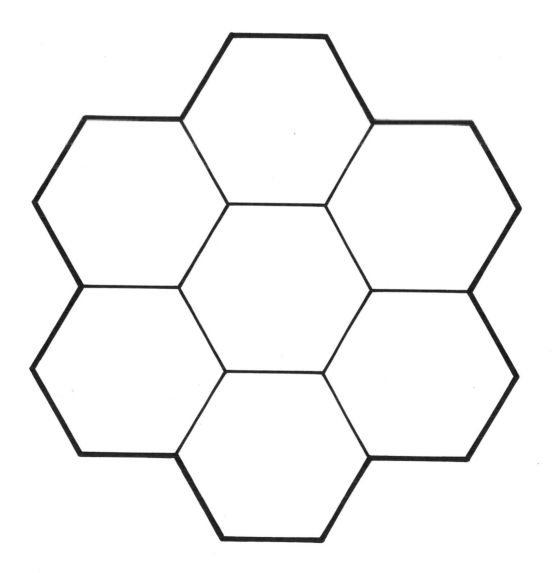

Cover with 4 blocks of one color.

Each block is _____ of _____ equal parts of the figure.

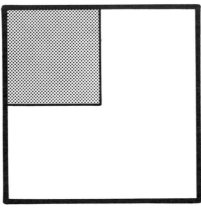

Cover $\frac{1}{4}$.

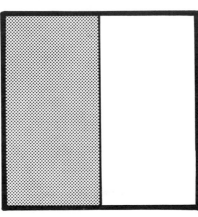

Cover $\frac{2}{4}$.

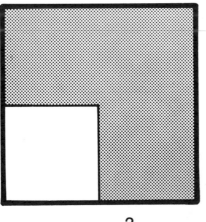

Cover $\frac{3}{4}$.

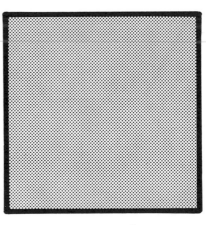

Cover $\frac{4}{4}$.

Write the fraction numerals:

one-fourth _____

three-fourths _____

two-fourths _____

four-fourths _____

24

Cover 3 of 4 equal parts of each figure.

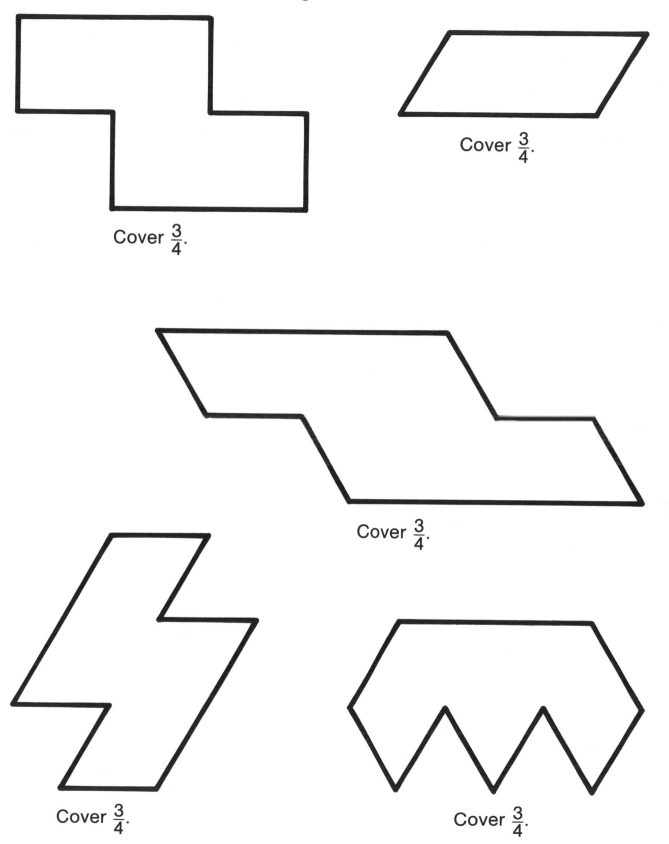

Cover $\frac{3}{4}$.

Cover $\frac{3}{4}$.

Cover $\frac{3}{4}$.

Cover $\frac{3}{4}$.

Cover $\frac{3}{4}$.

25

1) Cover each figure with 4 blocks of one color.

2) Trace around the 4 blocks.

3) Cover the figure again with blue and green blocks as directed.

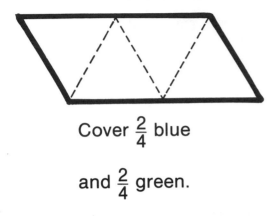

Cover $\frac{2}{4}$ blue

and $\frac{2}{4}$ green.

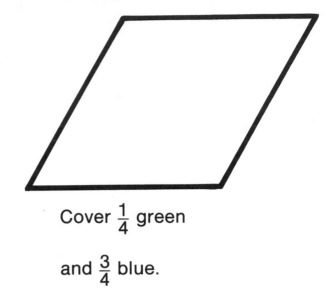

Cover $\frac{1}{4}$ green

and $\frac{3}{4}$ blue.

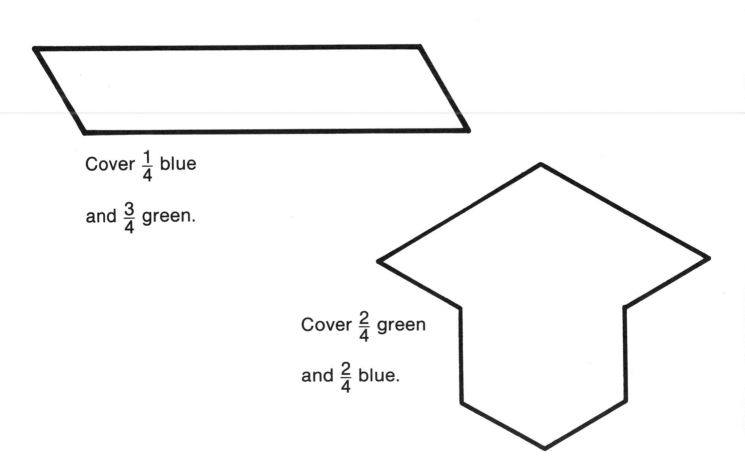

Cover $\frac{1}{4}$ blue

and $\frac{3}{4}$ green.

Cover $\frac{2}{4}$ green

and $\frac{2}{4}$ blue.

26

Cover with blue blocks. Each block is _____ of _____ equal parts of the figure. We call that one-fifth and write it $\frac{1}{5}$.

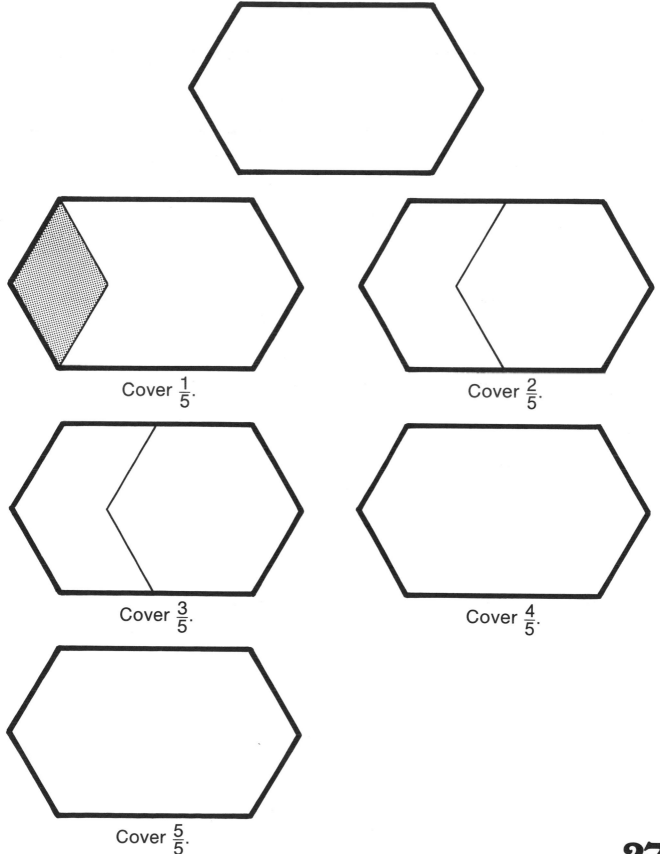

Cover $\frac{1}{5}$.

Cover $\frac{2}{5}$.

Cover $\frac{3}{5}$.

Cover $\frac{4}{5}$.

Cover $\frac{5}{5}$.

27

Cover with pattern blocks.

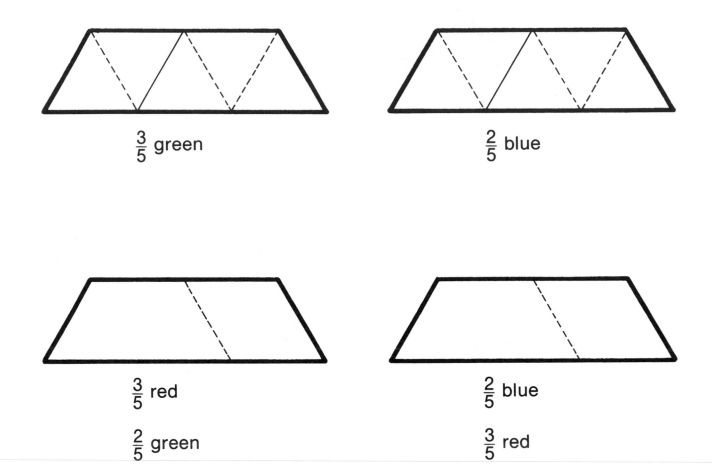

$\frac{3}{5}$ green

$\frac{2}{5}$ blue

$\frac{3}{5}$ red

$\frac{2}{5}$ green

$\frac{2}{5}$ blue

$\frac{3}{5}$ red

Build a design out of 5 red blocks. Trace it and shade $\frac{2}{5}$.

28

Cover with pattern blocks.

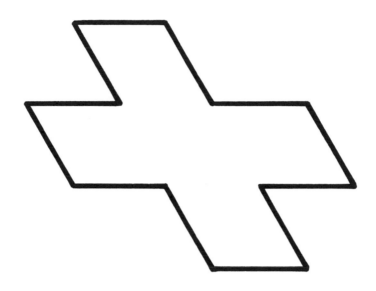

1) $\frac{5}{5}$ blue

2) $\frac{4}{5}$ blue

    $\frac{1}{5}$ green

3) $\frac{3}{5}$ red

    $\frac{2}{5}$ blue

4) $\frac{3}{5}$ red

    $\frac{1}{5}$ blue

    $\frac{1}{5}$ green

5) If you answered 3 of the 5 questions on this page, what fraction of the questions have you answered?

First, cover each figure below with one color block.

Next, cover $\frac{3}{5}$ of each figure below.

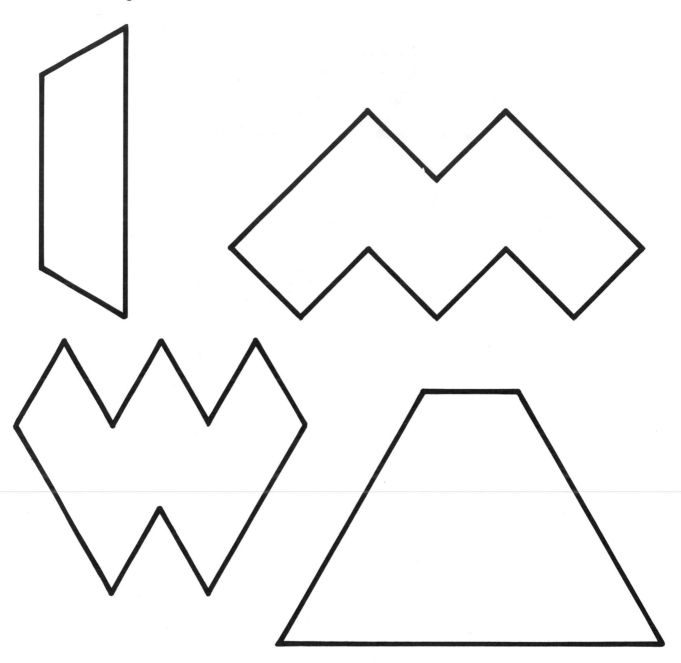

In $\frac{3}{5}$, what does the 5 mean? _____

_____

In $\frac{3}{5}$, what does the 3 mean? _____

_____

30

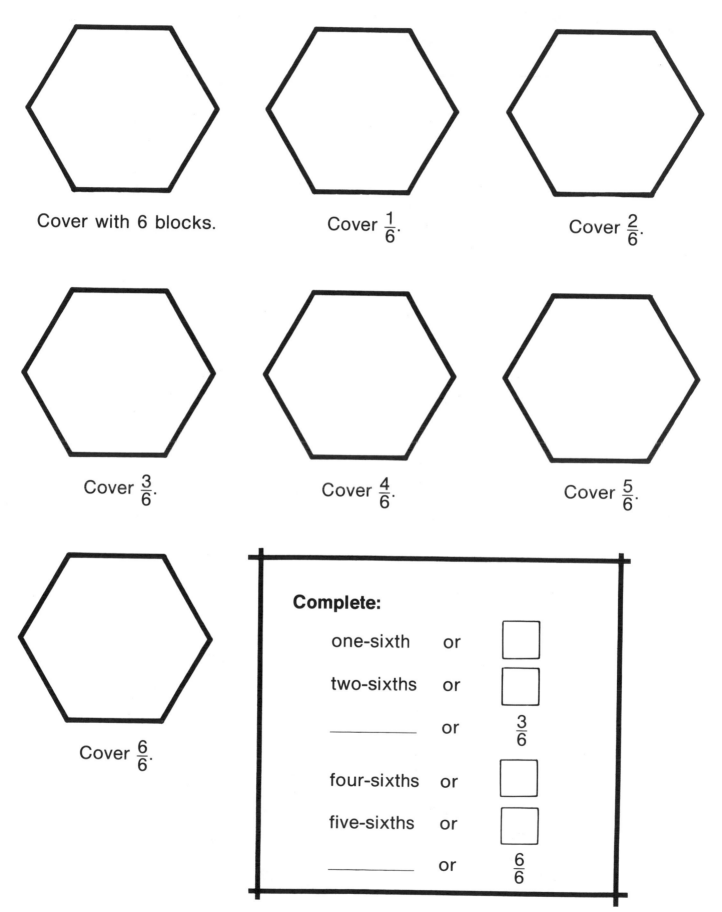

Cover with 6 blocks.

Cover $\frac{1}{6}$.

Cover $\frac{2}{6}$.

Cover $\frac{3}{6}$.

Cover $\frac{4}{6}$.

Cover $\frac{5}{6}$.

Cover $\frac{6}{6}$.

Complete:

one-sixth   or   ☐

two-sixths   or   ☐

_____   or $\frac{3}{6}$

four-sixths   or   ☐

five-sixths   or   ☐

_____   or $\frac{6}{6}$

31

Cover:

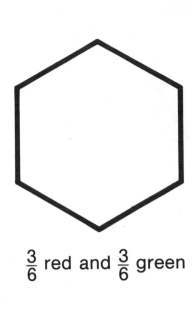

$\frac{3}{6}$ red and $\frac{3}{6}$ green

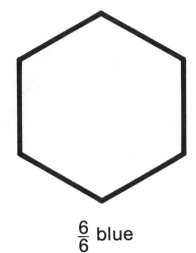

$\frac{6}{6}$ blue

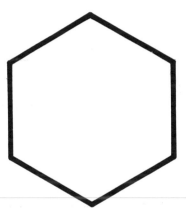

$\frac{4}{6}$ blue and $\frac{2}{6}$ green

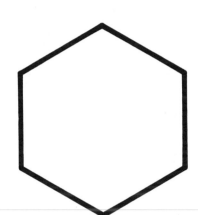

$\frac{6}{6}$ yellow

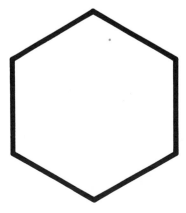

$\frac{2}{6}$ blue and $\frac{4}{6}$ green

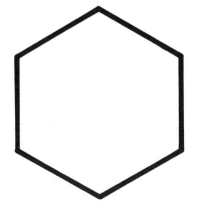

$\frac{1}{6}$ green, $\frac{2}{6}$ blue and $\frac{3}{6}$ red

32

Use pattern blocks to solve the 5 puzzles below:

1) Cover with blue blocks.

2) Cover with green blocks.

3) Cover $\frac{3}{6}$ blue and $\frac{3}{6}$ green.

4) Cover $\frac{3}{6}$ yellow and $\frac{3}{6}$ green.

5) Cover $\frac{1}{6}$ green, $\frac{2}{6}$ blue and $\frac{3}{6}$ red.

Circle the correct answer to each question.

Which is more? $\frac{1}{3}$ $\frac{2}{3}$

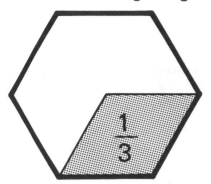   or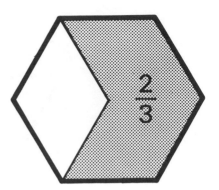

Which is more? $\frac{1}{6}$ $\frac{2}{6}$ $\frac{4}{6}$

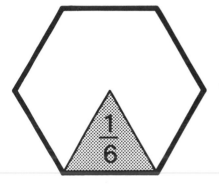   or   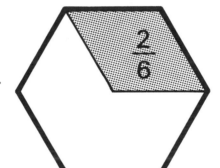   or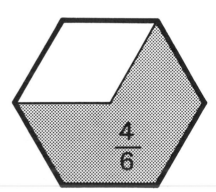

Which is more? $\frac{3}{5}$ $\frac{5}{10}$

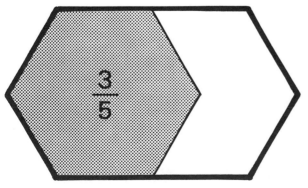   or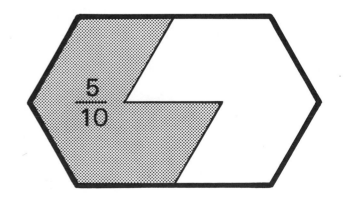

34

Circle the correct answer to each question.

Which is more? $\frac{2}{6}$ $\frac{1}{2}$ $\frac{2}{3}$

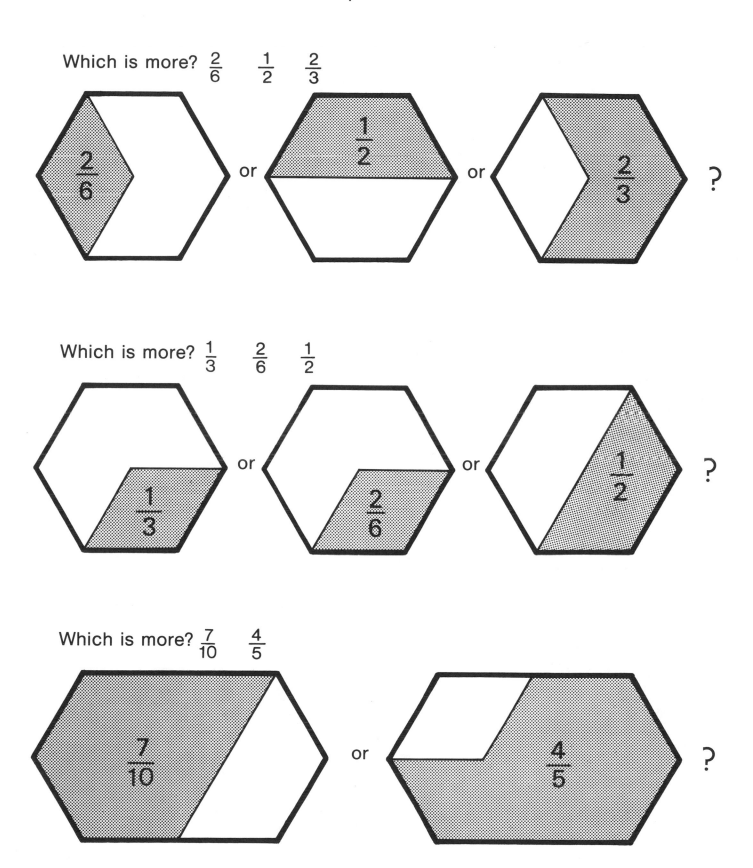

or   or   ?

Which is more? $\frac{1}{3}$ $\frac{2}{6}$ $\frac{1}{2}$

or   or   ?

Which is more? $\frac{7}{10}$ $\frac{4}{5}$

or   ?

35

One-half of each figure below is shaded.
Circle the answer.

Which shaded part needs more wood to cover? A B

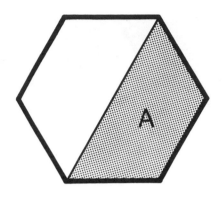
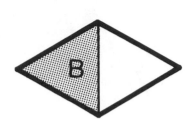

Which shaded part needs more wood to cover? C D

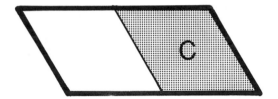
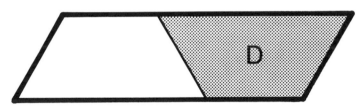

Which shaded part needs more wood to cover? E F

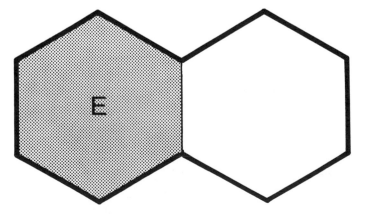
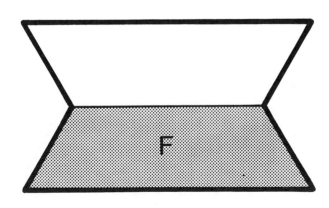

Is it true that $\frac{1}{2}$ of one thing may be bigger than $\frac{1}{2}$ of another? _____

36

Use pattern blocks to cover:

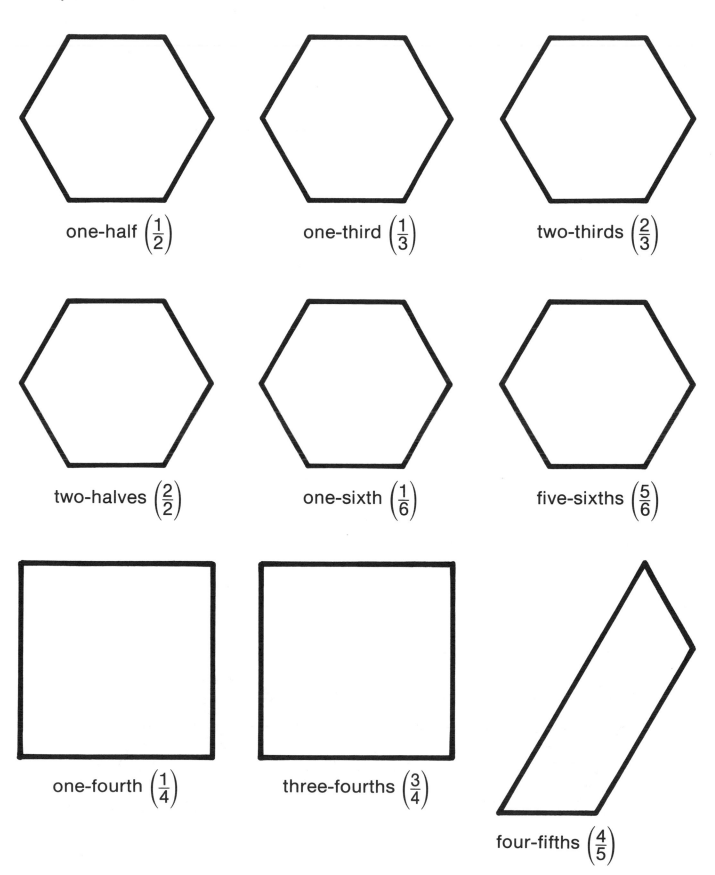

one-half $\left(\frac{1}{2}\right)$

one-third $\left(\frac{1}{3}\right)$

two-thirds $\left(\frac{2}{3}\right)$

two-halves $\left(\frac{2}{2}\right)$

one-sixth $\left(\frac{1}{6}\right)$

five-sixths $\left(\frac{5}{6}\right)$

one-fourth $\left(\frac{1}{4}\right)$

three-fourths $\left(\frac{3}{4}\right)$

four-fifths $\left(\frac{4}{5}\right)$

37

Cover the whole figure with green triangles.

It takes _____ triangles to cover the figure.

What fraction of the figure is shaded? ☐

Cover the whole figure with green triangles. It takes _____ triangles to cover the figure.

What fraction of the figure is shaded? ☐

Cover the whole figure with blue parallelograms.

It takes _____ parallelograms to cover the figure.

What fraction of the whole figure is shaded? ☐

38

How many green triangles cover this figure? _____

What fraction of the figure is shaded? ☐

How many orange squares cover this figure? _____

What fraction of the figure is shaded? ☐

How many green triangles cover this figure? _____

What fraction of the figure is shaded? ☐

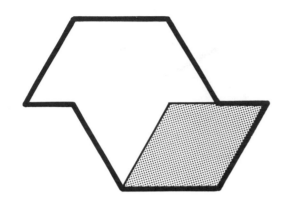

Cover the figure with blocks of one color.

How many blocks did it take?

_____

What fraction of the figure is shaded? ☐

Cover the figure with blocks of one color.

How many blocks did it take?

_____

What fraction of the figure is shaded? ☐

40

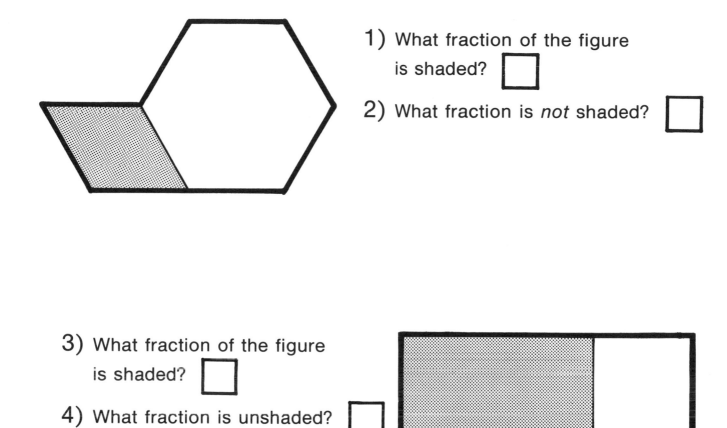

1) What fraction of the figure is shaded? ☐

2) What fraction is *not* shaded? ☐

3) What fraction of the figure is shaded? ☐

4) What fraction is unshaded? ☐

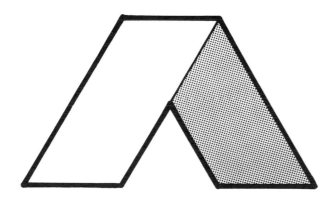

5) What fraction of the figure is shaded? ☐

6) What fraction of the figure is unshaded? ☐

41

A fraction may have many different names. Below are shown some different names for one-half $\left(\frac{1}{2}\right)$.

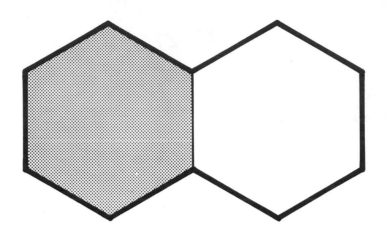

1) Covering with yellow hexagons _____half is shaded.

2) Covering with red trapezoids _____fourths are shaded.

3) Covering with blue parallelograms _____sixths are shaded.

4) Covering with green triangles _____twelfths are shaded.

5) What are some of the names for $\frac{1}{2}$? □ □ □ □

42

What fraction of the figure is shaded?

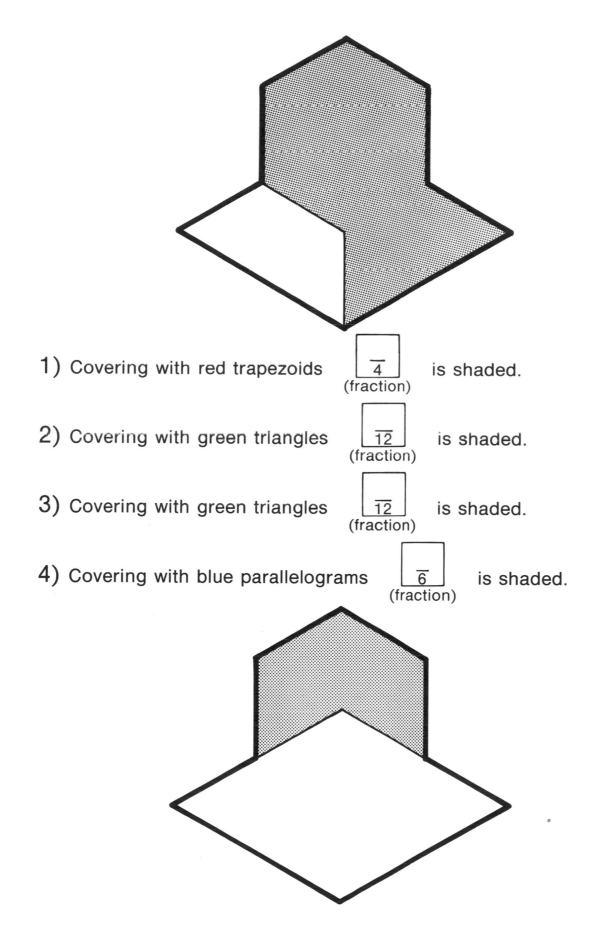

1) Covering with red trapezoids $\dfrac{}{4}$ (fraction) is shaded.

2) Covering with green triangles $\dfrac{}{12}$ (fraction) is shaded.

3) Covering with green triangles $\dfrac{}{12}$ (fraction) is shaded.

4) Covering with blue parallelograms $\dfrac{}{6}$ (fraction) is shaded.

What fraction of the figure is shaded?

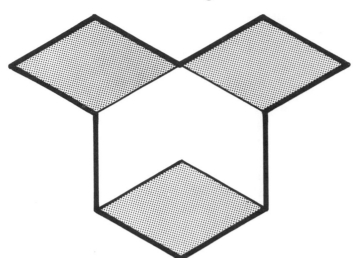

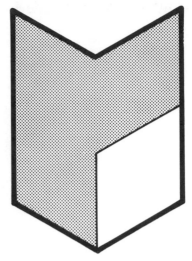

1) Covering with blue
   parallelograms ☐
   is shaded.

2) Covering with green
   triangles ☐ is
   shaded.

3) Covering with blue
   parallelograms ☐
   is shaded.

4) Covering with green
   triangles ☐ is
   shaded.

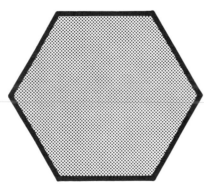

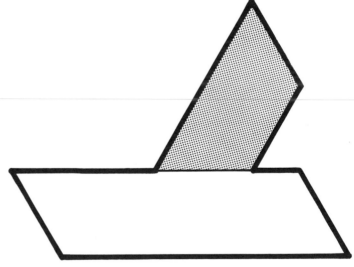

5) Covering with green
   triangles ☐ is
   shaded.

6) Covering with blue
   parallelograms ☐
   is shaded.

7) Covering with red
   trapezoids ☐
   is shaded.

8) Covering with red
   trapezoids ☐
   is shaded.

9) Covering with green
   triangles ☐ is
   shaded.

44

Build this figure with 2 red and 2 blue pattern blocks.

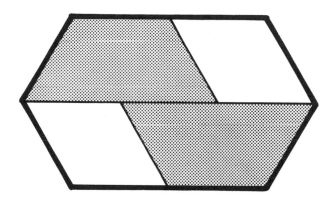

1) Are there the same number of red blocks as blue blocks?

_____

2) Is there the same amount of red wood as blue wood? _____

3) How many green triangles cover the blue wood? _____

4) How many green triangles cover the red wood? _____

5) How many green triangles cover the whole figure? _____

6) What fraction of the figure is blue? ☐

7) What fraction is red? ☐

8) Which fraction is greater, $\frac{6}{10}$ or $\frac{4}{10}$? ☐

45

Build the figure below with pattern blocks.

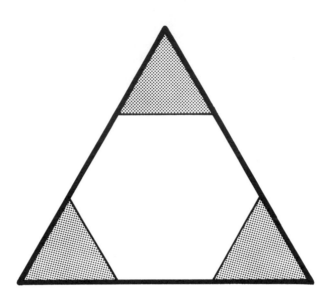

1) Are there more triangles than hexagons? _____

2) Does that mean the triangles form a larger part of the figure?

_____

Now make the figure using all triangles.

3) How many triangles cover the whole figure? _____

4) What fraction of figure above is shaded? ☐

5) What fraction is not shaded? ☐

6) Is $\frac{3}{9}$ larger than $\frac{6}{9}$? _____

46

Build this figure with yellow and green pattern blocks.

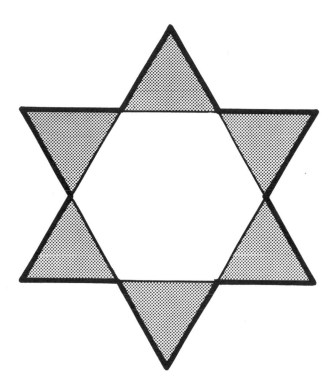

1) Is there more green wood or more yellow wood? _____

2) What fraction is yellow? ⬜

3) What fraction is green? ⬜

Build this figure with red and green pattern blocks.

4) Is there more red wood or more green wood? _____

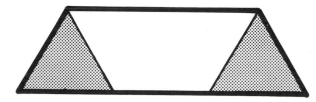

5) What fraction is red? ⬜

6) What fraction is green? ⬜

7) Circle the larger fraction: $\frac{2}{5}$ $\frac{3}{5}$

47

Build this figure with red and yellow pattern blocks.

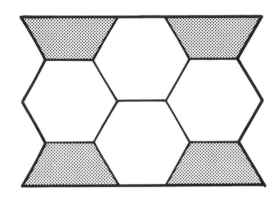

1) Is there more red wood or more yellow wood? _____

2) How many red blocks would cover the entire figure?

   _____

3) What fraction of the figure is red? ☐

4) What fraction is yellow? ☐

5) Fill in the boxes with the two fractions.

   ☐   ☐
   (fraction)   is greater than   (fraction)

Build this figure with red and yellow pattern blocks.

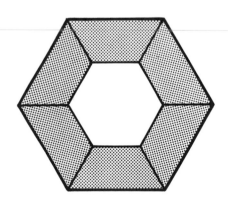

6) Is there more red wood or more yellow wood? _____

7) What fraction is red? ☐

8) What fraction is yellow? ☐

9) Fill in the boxes with the two fractions.

   ☐   ☐
   (fraction)   is greater than   (fraction)

48

Use pattern blocks to help answer these questions.

Is more of the figure shaded or more unshaded? Circle the answer.

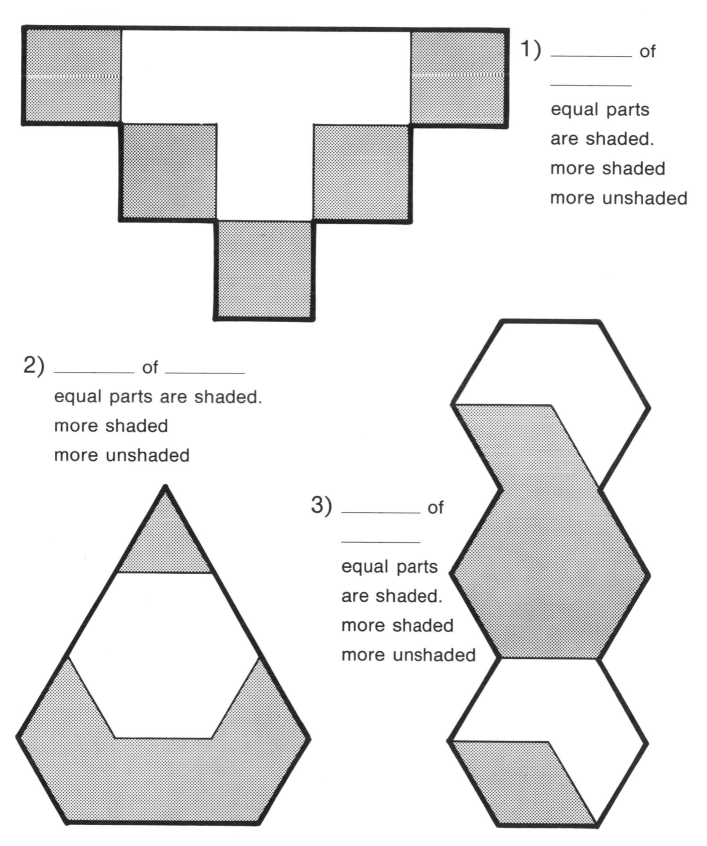

1) _____ of

_____

equal parts

are shaded.

more shaded

more unshaded

2) _____ of _____

equal parts are shaded.

more shaded

more unshaded

3) _____ of

_____

equal parts

are shaded.

more shaded

more unshaded

49

Is more of the figure shaded or unshaded? Circle the answer.

1)

more shaded

more unshaded

2)

more shaded

more unshaded

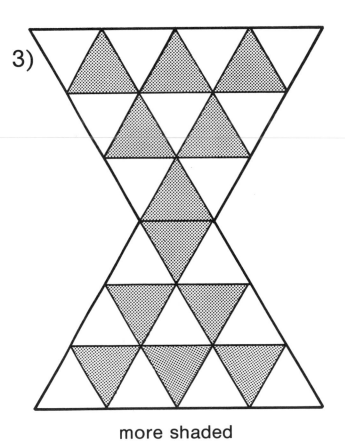

3)

more shaded

more unshaded

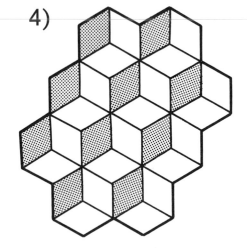

4)

more shaded

more unshaded

50

Is more of the figure shaded or unshaded? Circle the answer.

1)

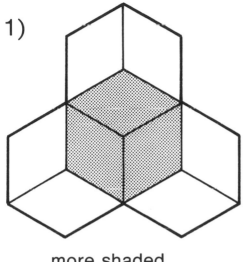

more shaded

more unshaded

2)

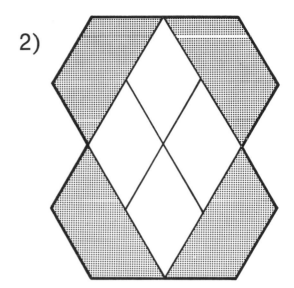

more shaded

more unshaded

3)

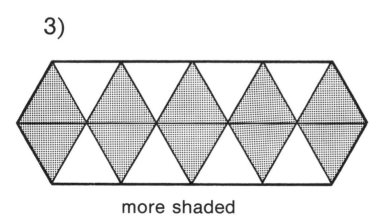

more shaded

more unshaded

4)

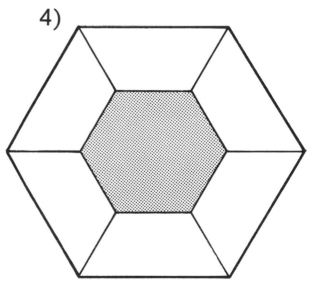

more shaded

more unshaded

Use pattern blocks to solve these problems.

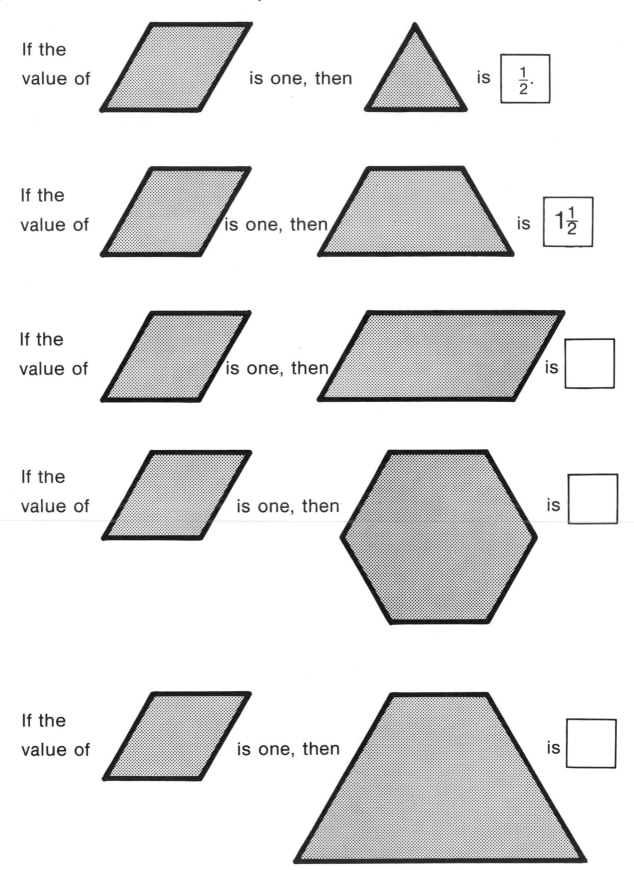

If the value of ▱ is one, then △ is $\boxed{\frac{1}{2}}$.

If the value of ▱ is one, then ⬡ is $\boxed{1\frac{1}{2}}$

If the value of ▱ is one, then ▱ is $\boxed{}$

If the value of ▱ is one, then ⬡ is $\boxed{}$

If the value of ▱ is one, then ⬡ is $\boxed{}$

52

Use pattern blocks to solve these problems.

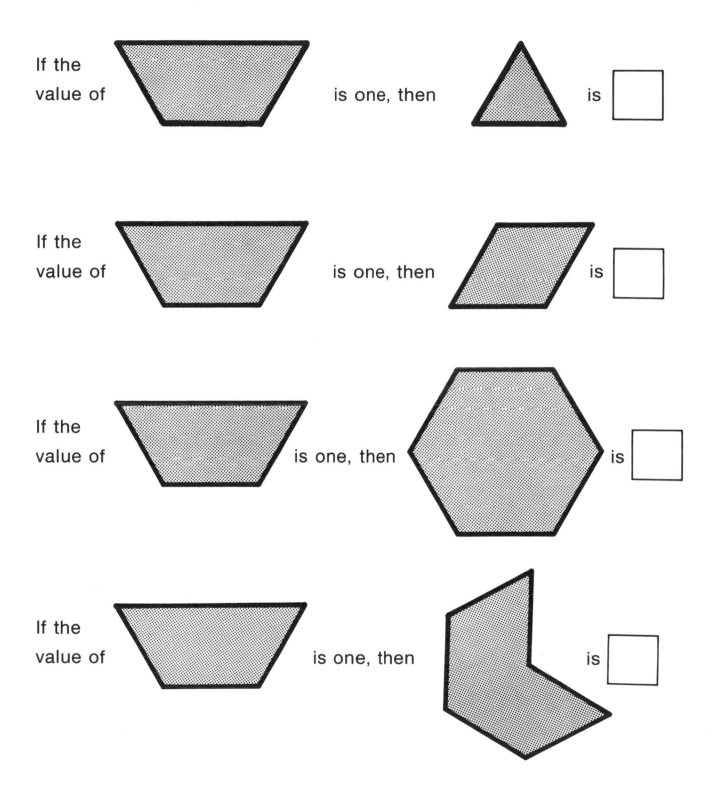

Use pattern blocks to solve these problems.

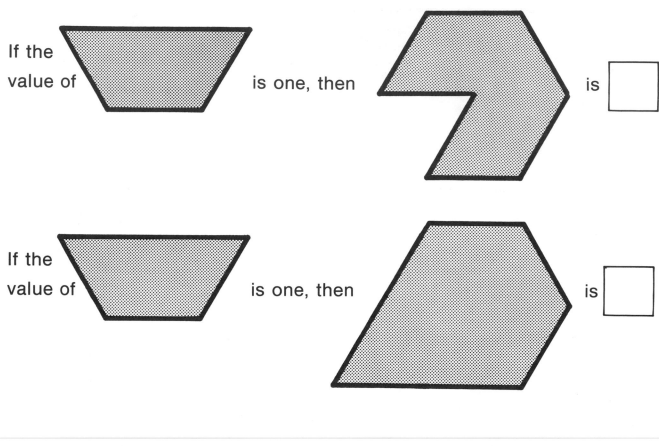

If the value of [trapezoid] is one, then [shape] is [ ]

If the value of [trapezoid] is one, then [shape] is [ ]

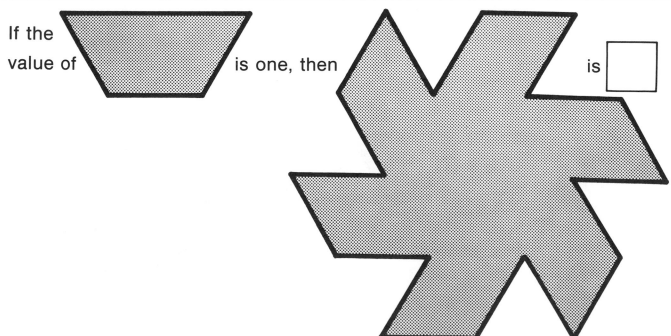

If the value of [trapezoid] is one, then [shape] is [ ]

Draw your answers by tracing the blocks or by using a pattern block template.

If 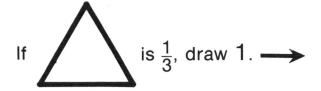 is $\frac{1}{3}$, draw 1. ⟶

If 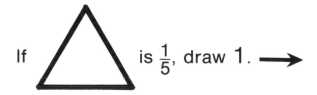 is $\frac{1}{5}$, draw 1. ⟶

If △ is $\frac{1}{2}$, draw 1. ⟶

If △ is $\frac{1}{4}$, draw 1. ⟶

Trace pattern blocks or use a template.

If 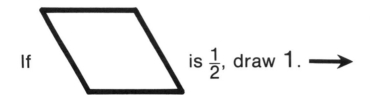 is $\frac{1}{2}$, draw 1. →

If is $\frac{1}{3}$, draw 1. →

If is $\frac{1}{2}$, draw $1\frac{1}{2}$. →

Draw the answers to these problems on a separate sheet of paper. Trace the pattern blocks or use a pattern block template.

1) If 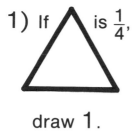 is $\frac{1}{4}$,

draw 1.

2) If 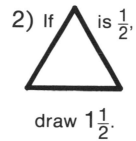 is $\frac{1}{2}$,

draw $1\frac{1}{2}$.

3) If 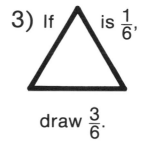 is $\frac{1}{6}$,

draw $\frac{3}{6}$.

4) If 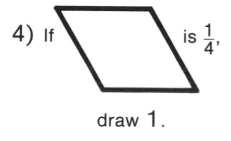 is $\frac{1}{4}$,

draw 1.

5) If 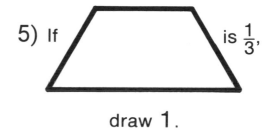 is $\frac{1}{3}$,

draw 1.

6) If 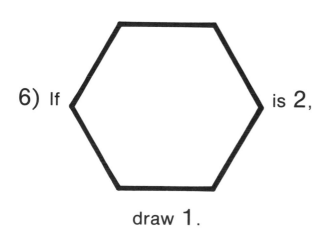 is 2,

draw 1.

7) If 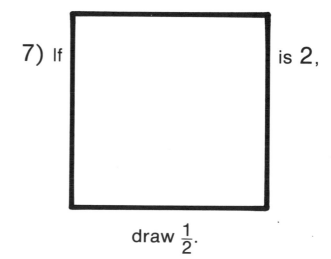 is 2,

draw $\frac{1}{2}$.

57

THE BLOCK EXCHANGE

Number of players: 3–5

Materials: Pattern block hexagons, trapezoids, blue parallelograms, and triangles. One die or a 1–6 spinner.

Rules:

1) Choose one player to be the broker.

2) In turn, each player rolls the die (or spins) and receives from the broker the number of triangles indicated on the die (or spinner).

3) The player then trades for equivalent larger blocks. Each player should have the fewest number of blocks possible at all times.
   Example: A player rolls 5, receives 5 triangles from the broker, and trades for one trapezoid and one parallelogram.

4) The winner is the first player to accumulate 5 hexagons.

In the activities that follow we will be using the four pattern blocks below.

1) Cover the hexagon above with triangles.

2) Cover the hexagon with parallelograms.

3) Cover the hexagon with trapezoids.

4) Cover the trapezoid with triangles.

5) Cover the parallelogram with triangles.

59

In the activities that follow, this figure represents ONE whole.

Complete the sentences.

_____ ⬡ cover one whole ⬡⬡ .

_____ ⬟ cover one whole ⬡⬡ .

_____ ▱ cover one whole ⬡⬡ .

_____ △ cover one whole ⬡⬡ .

60

Cover the figure below with red trapezoids.

How many did you use? _____

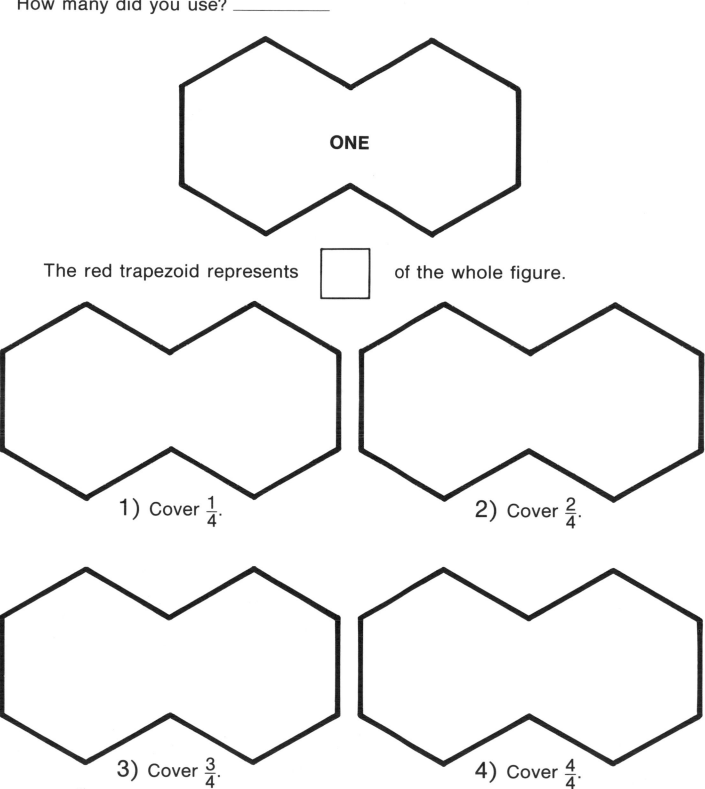

ONE

The red trapezoid represents ▢ of the whole figure.

1) Cover $\frac{1}{4}$.

2) Cover $\frac{2}{4}$.

3) Cover $\frac{3}{4}$.

4) Cover $\frac{4}{4}$.

Addition of fractions involves combining parts. With pattern blocks this means putting together blocks.

$$\frac{1}{4} + \frac{1}{4} = \frac{2}{4}$$

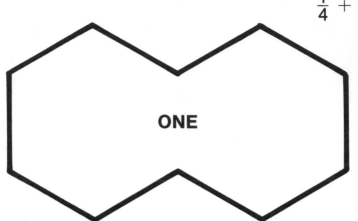

ONE

Use red trapezoids to solve each problem.

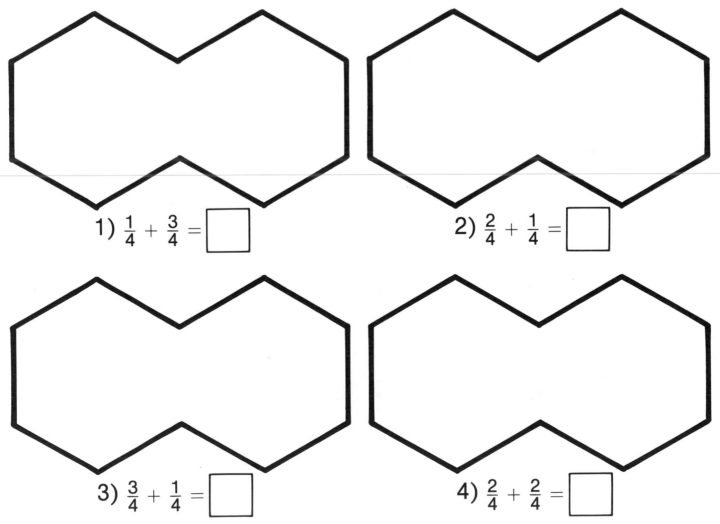

1) $\frac{1}{4} + \frac{3}{4} = \boxed{}$

2) $\frac{2}{4} + \frac{1}{4} = \boxed{}$

3) $\frac{3}{4} + \frac{1}{4} = \boxed{}$

4) $\frac{2}{4} + \frac{2}{4} = \boxed{}$

62

Cover the figure below with blue parallelograms.

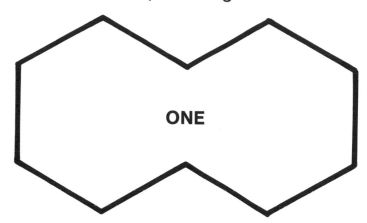

How many blue blocks did it take? _____

Each blue block covers $\frac{1}{6}$ of the whole figure.

Use blue pattern blocks to cover:

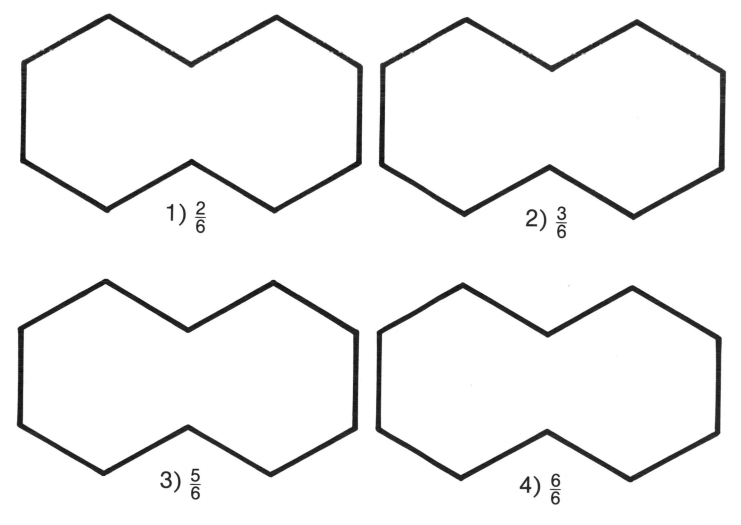

1) $\frac{2}{6}$

2) $\frac{3}{6}$

3) $\frac{5}{6}$

4) $\frac{6}{6}$

63

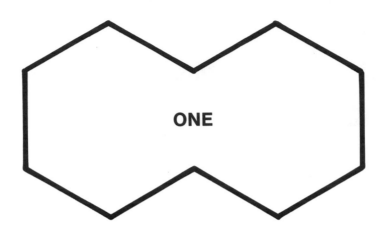

ONE

Use blue parallelograms to solve these problems.

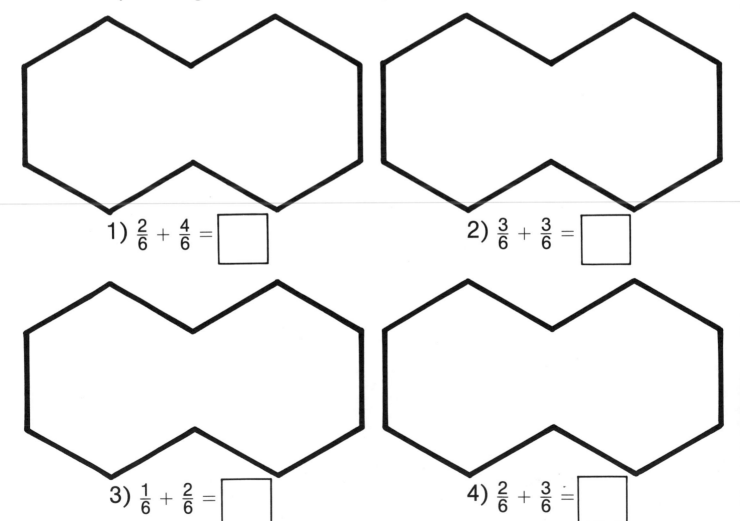

1) $\frac{2}{6} + \frac{4}{6} =$ ☐

2) $\frac{3}{6} + \frac{3}{6} =$ ☐

3) $\frac{1}{6} + \frac{2}{6} =$ ☐

4) $\frac{2}{6} + \frac{3}{6} =$ ☐

64

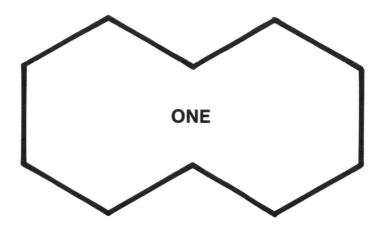

ONE

Use pattern blocks to solve these problems.

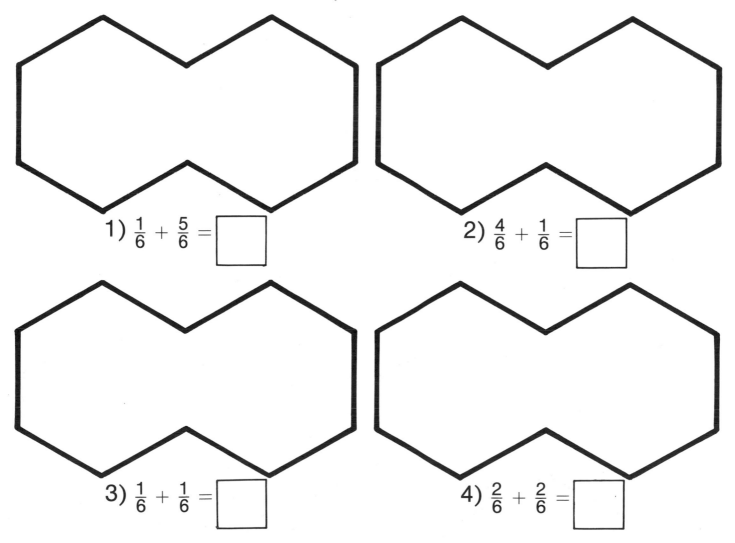

1) $\frac{1}{6} + \frac{5}{6} = \boxed{}$

2) $\frac{4}{6} + \frac{1}{6} = \boxed{}$

3) $\frac{1}{6} + \frac{1}{6} = \boxed{}$

4) $\frac{2}{6} + \frac{2}{6} = \boxed{}$

65

Use blocks to solve each problem.

ONE

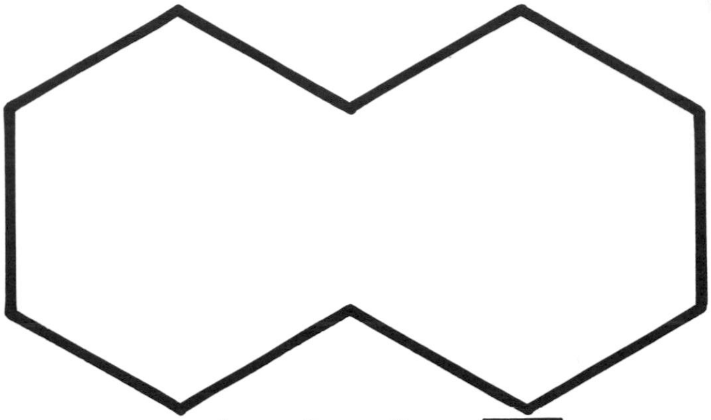

1) $\frac{1}{6} + \frac{2}{6} + \frac{3}{6} = \square$

2) $\frac{1}{6} + \frac{1}{6} + \frac{1}{6} = \square$

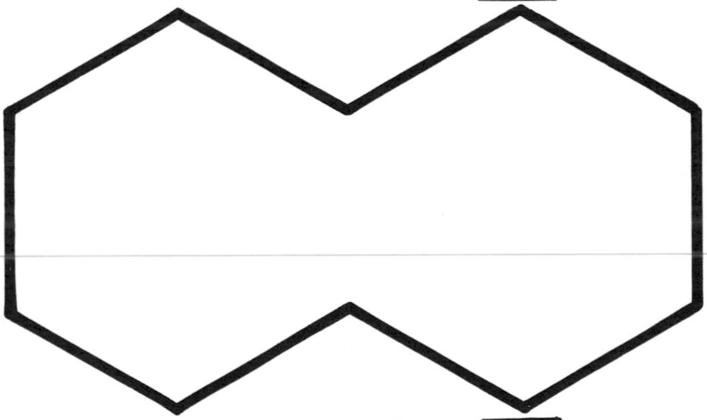

3) $\frac{1}{6} + \frac{4}{6} + \frac{1}{6} = \square$

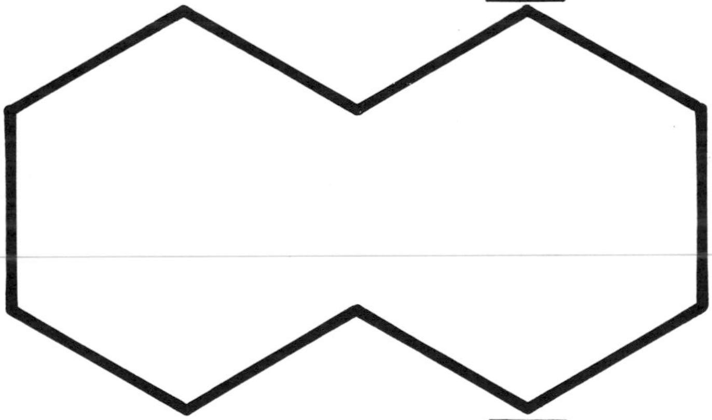

4) $\frac{2}{6} + \frac{1}{6} + \frac{2}{6} = \square$

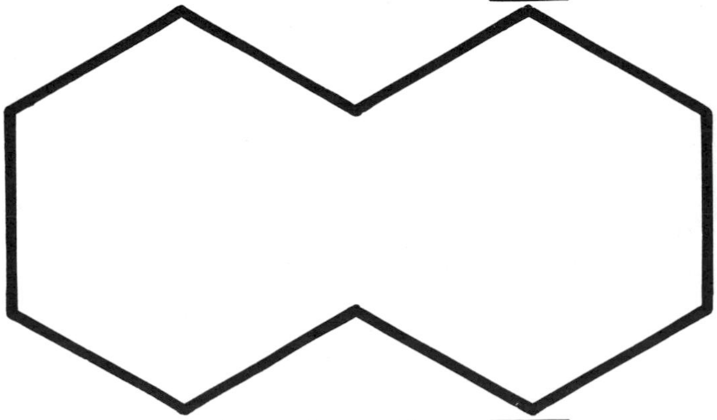

5) $\frac{2}{6} + \frac{2}{6} + \frac{2}{6} = \square$

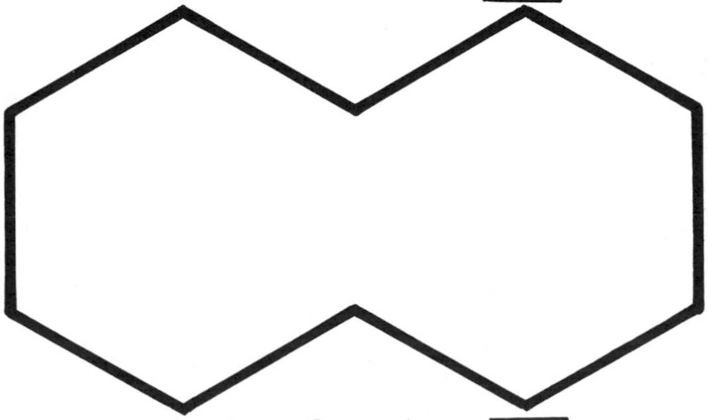

6) $\frac{1}{6} + \frac{3}{6} + \frac{1}{6} = \square$

66

Cover the figure below with green triangles.

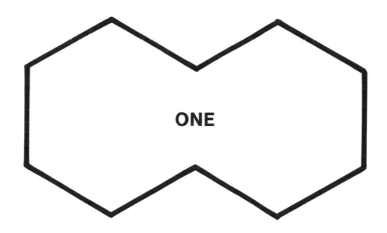

How many did you use? _____

Each triangle covers ☐ of the figure.

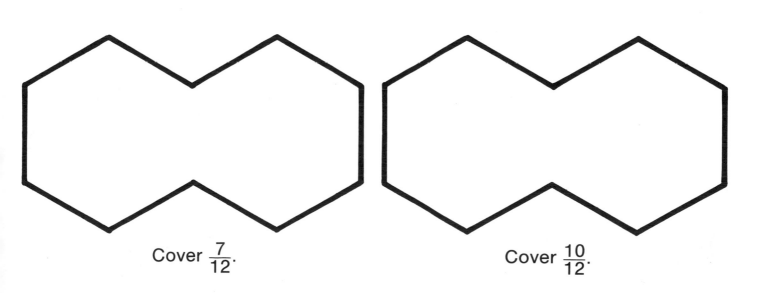

Cover $\frac{7}{12}$.

Cover $\frac{10}{12}$.

Use blocks to
solve each problem.

ONE

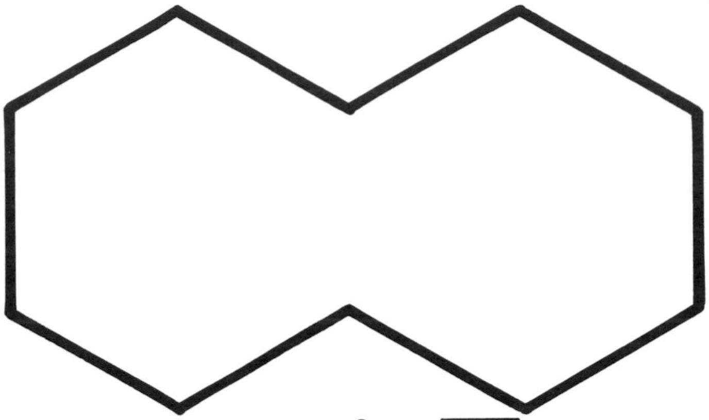

1) $\frac{6}{12} + \frac{6}{12} = \square$

2) $\frac{5}{12} + \frac{7}{12} = \square$

3) $\frac{3}{12} + \frac{3}{12} = \square$

4) $\frac{1}{12} + \frac{5}{12} = \square$

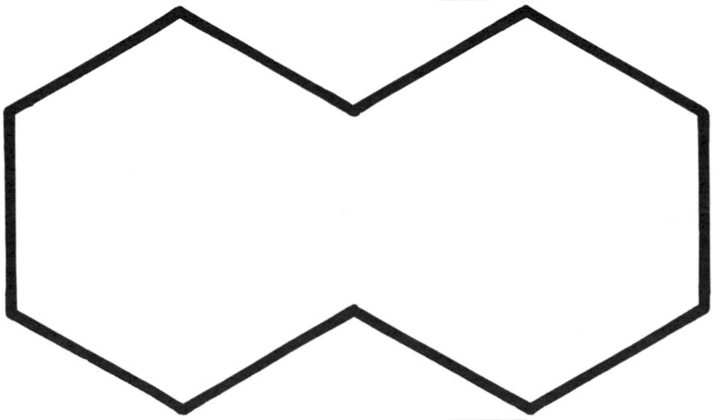

5) $\frac{4}{12} + \frac{3}{12} = \square$

6) $\frac{11}{12} + \frac{1}{12} = \square$

68

Use blocks to solve each problem.

Adding twelfths

ONE

1) $\frac{4}{12} + \frac{4}{12} + \frac{4}{12} = \square$

2) $\frac{4}{12} + \frac{3}{12} + \frac{5}{12} = \square$

3) $\frac{9}{12} + \frac{1}{12} + \frac{1}{12} = \square$

4) $\frac{3}{12} + \frac{2}{12} + \frac{1}{12} = \square$

5) $\frac{6}{12} + \frac{2}{12} + \frac{1}{12} = \square$

6) $\frac{1}{12} + \frac{7}{12} + \frac{3}{12} = \square$

69

In a fraction the top number is called the *numerator* and the bottom number is called the *denominator*.

$$\frac{3}{4}$$

3 ← NUMERATOR

4 ← DENOMINATOR

Fractions which have the same denominator are said to have *common denominators* or *like denominators*.

Circle the problems below which contain common denominators.

1) $\frac{1}{6} + \frac{3}{6}$

2) $\frac{1}{4} + \frac{1}{4}$

3) $\frac{1}{2} + \frac{2}{2}$

4) $\frac{3}{4} + \frac{3}{6}$

5) $\frac{1}{5} + \frac{5}{12}$

6) $\frac{5}{12} + \frac{5}{12}$

7) $\frac{1}{10} + \frac{3}{10} + \frac{5}{10}$

8) $\frac{1}{8} + \frac{3}{8} + \frac{5}{8}$

9) $\frac{4}{5} + \frac{4}{6}$

70

Do you see a pattern in the addition of fractions with common denominators?

Adding fourths to fourths gives an answer in fourths.

$$\frac{?}{4} + \frac{?}{4} = \frac{?}{4}$$

Adding sixths to sixths gives an answer in sixths.

$$\frac{?}{6} + \frac{?}{6} = \frac{?}{6}$$

First, solve these problems *without* pattern blocks. Next, use the blocks to check your answers.

1) $\frac{1}{6} + \frac{1}{6} = \boxed{}$ 2) $\frac{5}{12} + \frac{1}{12} = \boxed{}$

3) $\frac{1}{4} + \frac{2}{4} = \boxed{}$ 4) $\frac{3}{12} + \frac{7}{12} = \boxed{}$

5) $\frac{6}{12} + \frac{4}{12} = \boxed{}$ 6) $\frac{3}{6} + \frac{2}{6} = \boxed{}$

Describe a pattern you found in solving these addition problems.

_____

_____

Fractions which are the same part of a whole are called *equivalent* fractions.

Cover the figures below with pattern blocks. Show that the fractions are equivalent.

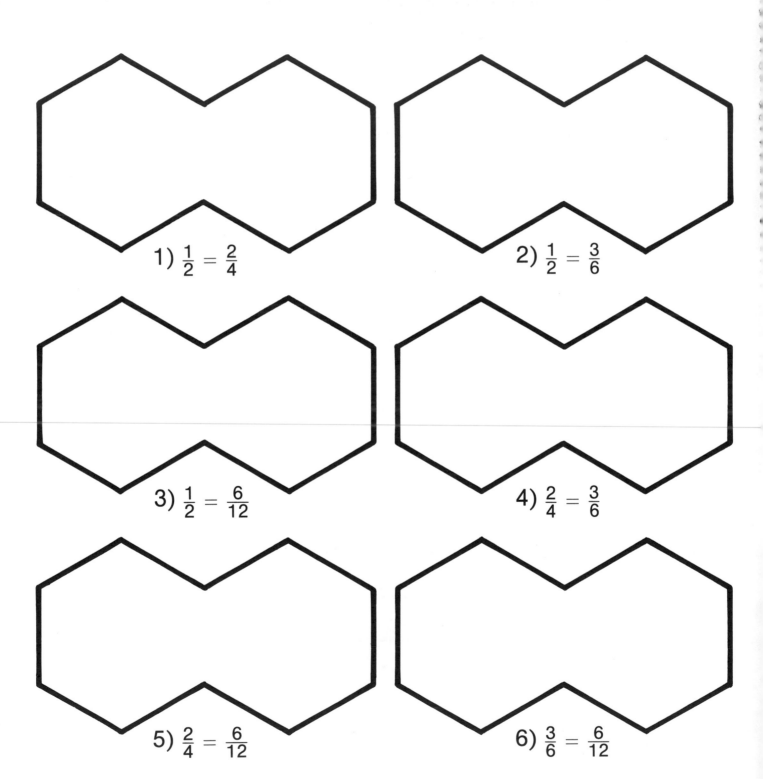

1) $\frac{1}{2} = \frac{2}{4}$

2) $\frac{1}{2} = \frac{3}{6}$

3) $\frac{1}{2} = \frac{6}{12}$

4) $\frac{2}{4} = \frac{3}{6}$

5) $\frac{2}{4} = \frac{6}{12}$

6) $\frac{3}{6} = \frac{6}{12}$

Use pattern blocks to show equivalent fractions.

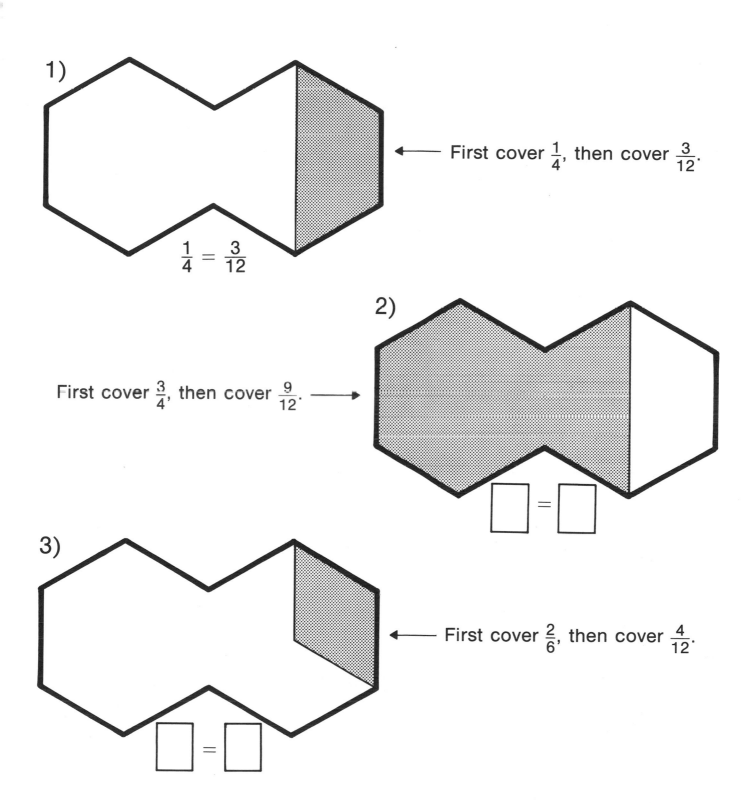

1)

First cover $\frac{1}{4}$, then cover $\frac{3}{12}$.

$\frac{1}{4} = \frac{3}{12}$

2)

First cover $\frac{3}{4}$, then cover $\frac{9}{12}$.

$\square = \square$

3)

First cover $\frac{2}{6}$, then cover $\frac{4}{12}$.

$\square = \square$

Use pattern blocks to show equivalent fractions.

1)

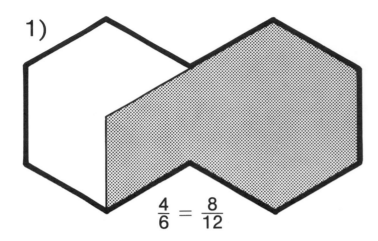

$$\frac{4}{6} = \frac{8}{12}$$

First cover $\frac{4}{6}$, then cover $\frac{8}{12}$.

Show that $\frac{4}{4}$ is equivalent to $\frac{6}{6}$

2)

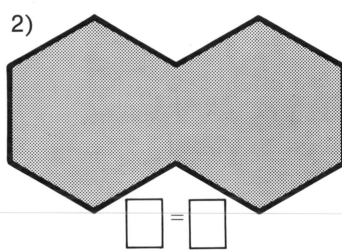

$\boxed{} = \boxed{}$

3)

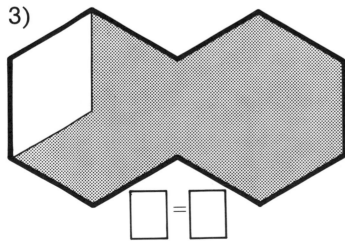

$\boxed{} = \boxed{}$

Show that $\frac{5}{6}$ is equivalent to $\frac{10}{12}$.

Are $\frac{3}{4}$ and $\frac{5}{6}$ equivalent fractions?

74

Look for patterns. Fill in the boxes.

1) $1 = \dfrac{1}{1} = \dfrac{2}{2} = \dfrac{3}{3} = \dfrac{\Box}{4} = \dfrac{\Box}{5} = \dfrac{\Box}{6} = \dfrac{\Box}{7} = \dfrac{8}{\Box}$

2) $\dfrac{1}{2} = \dfrac{2}{4} = \dfrac{3}{6} = \dfrac{4}{8} = \dfrac{5}{\Box} = \dfrac{6}{\Box} = \dfrac{7}{\Box} = \dfrac{\Box}{16} = \dfrac{\Box}{18}$

3) $\dfrac{1}{3} = \dfrac{2}{6} = \dfrac{3}{9} = \dfrac{4}{12} = \dfrac{\Box}{15} = \dfrac{\Box}{18} = \dfrac{\Box}{21} = \dfrac{8}{\Box}$

4) $\dfrac{1}{4} = \dfrac{2 \times 1}{2 \times 4} = \dfrac{2}{8} = \dfrac{3 \times 1}{3 \times 4} = \dfrac{3}{12} = \dfrac{4 \times 1}{4 \times 4} = \dfrac{\Box}{16} = \dfrac{5 \times 1}{\Box \times 4} = \dfrac{5}{20}$

5) $\dfrac{3}{4} = \dfrac{2 \times 3}{2 \times 4} = \dfrac{6}{8} = \dfrac{3 \times 3}{3 \times 4} = \dfrac{\Box}{12}$

6) $\dfrac{1}{6} = \dfrac{2}{12} = \dfrac{3}{18} = \dfrac{\Box}{24} = \dfrac{\Box}{30} = \dfrac{6}{\Box} = \dfrac{100}{\Box}$

75

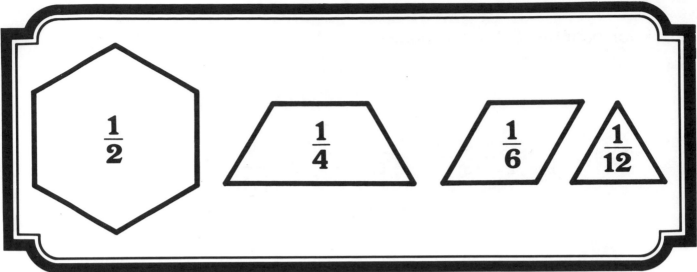

Use the values given above.

Trade one color wood for another.

Example:

Trade $\frac{3}{4}$ for green wood. How many twelfths? _____

 $\frac{3}{4} = \frac{9}{12}$

1) Trade $\frac{1}{2}$ for blue wood. How many sixths? _____

$$\frac{1}{2} = \boxed{}$$

2) Trade $\frac{4}{12}$ for blue wood. How many sixths? _____

$$\frac{4}{12} = \boxed{}$$

3) Trade $\frac{6}{12}$ for yellow wood. How many halves? _____

$$\frac{6}{12} = \boxed{}$$

4) Trade $\frac{5}{6}$ for green wood. How many twelfths? _____

$$\frac{5}{6} = \boxed{}$$

76

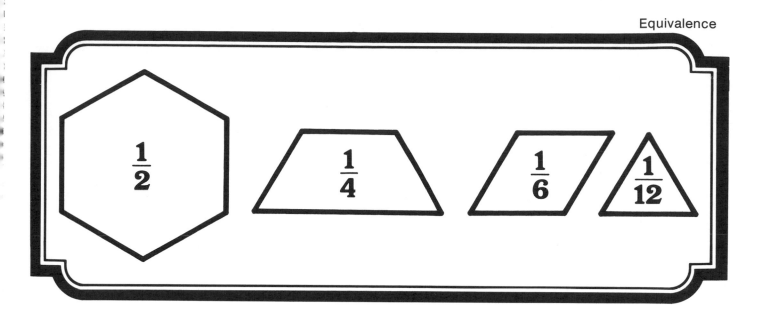

Trade the blocks for the *fewest* blocks of *one color.*

Example:

One yellow, one red. $\left(\frac{1}{2} + \frac{1}{4}\right)$ $\underline{\ \ \ 3\ \ \ }$ $\underline{\ \ \ red\ \ \ }$
(no.)   (color)

1) One yellow, one red, one blue $\left(\frac{1}{2} + \frac{1}{4} + \frac{1}{6}\right)$ $\underline{\hspace{2cm}}$
(no.)   (color)

2) One yellow, three blue $\left(\frac{1}{2} + \frac{3}{6}\right)$ $\underline{\hspace{2cm}}$
(no.)   (color)

3) Three red, three blue $\left(\frac{3}{4} + \frac{3}{6}\right)$ $\underline{\hspace{2cm}}$
(no.)   (color)

4) Four green, one blue $\left(\frac{4}{12} + \frac{1}{6}\right)$ $\underline{\hspace{2cm}}$
(no.)   (color)

5) One red, one blue, one green $\left(\frac{1}{4} + \frac{1}{6} + \frac{1}{12}\right)$ $\underline{\hspace{2cm}}$
(no.)   (color)

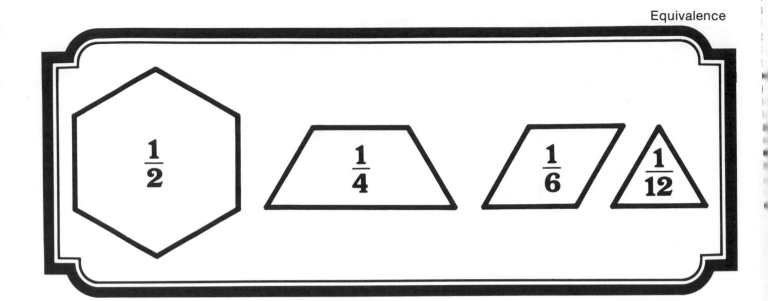

Trade the blocks for the *fewest* blocks of *one color.*

1) Five green, one blue $\left(\frac{5}{12} + \frac{1}{6}\right)$ _____
(no.)   (color)

2) Five blue, one red $\left(\frac{5}{6} + \frac{1}{4}\right)$ _____
(no.)   (color)

3) Two blue, two green, two yellow $\left(\frac{2}{6} + \frac{2}{12} + \frac{2}{2}\right)$ _____
(no.)   (color)

4) Seven green, one red $\left(\frac{7}{12} + \frac{1}{4}\right)$ _____
(no.)   (color)

5) Five blue, two green $\left(\frac{5}{6} + \frac{2}{12}\right)$ _____
(no.)   (color)

6) Nine green, three red $\left(\frac{9}{12} + \frac{3}{4}\right)$ _____
(no.)   (color)

78

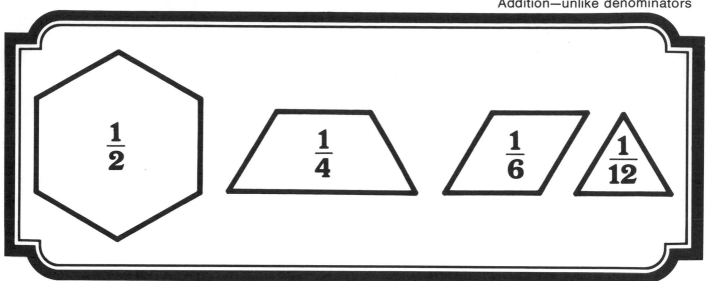

$\frac{1}{2}$ $\frac{1}{4}$ $\frac{1}{6}$ $\frac{1}{12}$

Trade the blocks for the *fewest* blocks of *one color*.

Write the answer as a fraction.

Example: $\frac{1}{6}$ + $\frac{1}{4}$ = $\frac{5}{12}$

             (1 blue + 1 red) = (5 green)

1) $\frac{1}{2}$ + $\frac{1}{4}$ =   ☐

     (1 yellow + 1 red) = (_____)
                          (no.)   (color)

2) $\frac{1}{2}$ + $\frac{1}{6}$ =   ☐

     (1 yellow + 1 blue) = (_____)
                          (no.)   (color)

3) $\frac{1}{2}$ + $\frac{1}{12}$ =   ☐

     (1 yellow + 1 green) = (_____)
                            (no.)   (color)

4) $\frac{1}{6}$ + $\frac{1}{12}$ =   ☐

     (1 blue + 1 green) = (_____)
                          (no.)   (color)

79

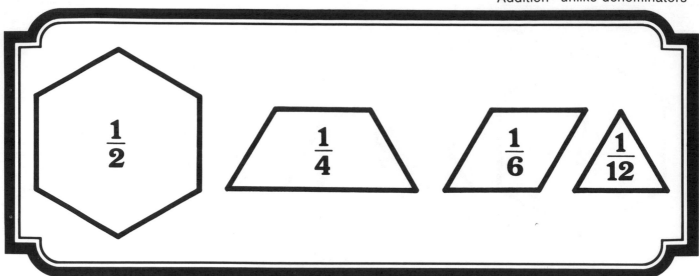

Trade the blocks for the *fewest* blocks of *one color*.

Solve these problems.

Example: $\frac{1}{12}$ + $\frac{1}{4}$ = $\frac{2}{6}$

(1 green + 1 red) = (2_____ blue)
                      (no.)   (color)

1) □ + □ = □

(1 yellow + 3 green) = (_____)
                        (no.)   (color)

2) □ + □ = □

(1 red + 2 blue) = (_____)
                    (no.)   (color)

3) □ + □ = □

(3 red + 1 blue) = (_____)
                    (no.)   (color)

4) □ + □ = □

(1 red + 5 green) = (_____)
                     (no.)   (color)

80

How can we describe the sum of $\frac{1}{4} + \frac{1}{6}$?

One way is to trade both blocks for the same color wood, that is, find a common denominator.

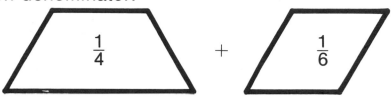

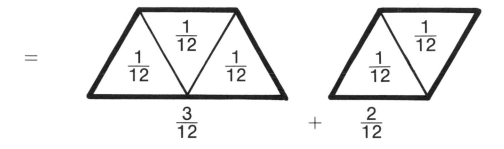

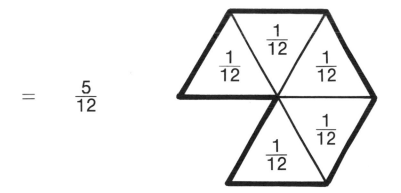

Use pattern blocks to change unlike denominators to common denominators (same color wood) and add the fractions.

1) $\frac{1}{12} + \frac{1}{2} = \boxed{\frac{}{12}} + \boxed{\frac{}{12}} = \boxed{}$

2) $\frac{1}{2} + \frac{1}{6} = \boxed{} + \boxed{} = \boxed{}$

81

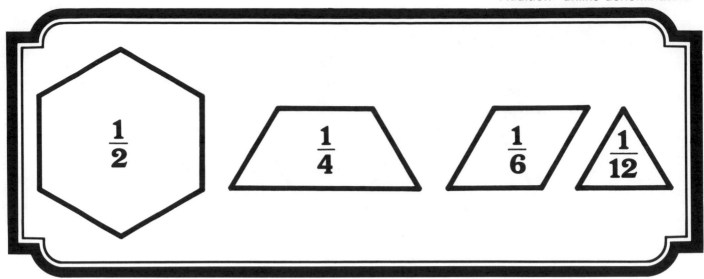

Change unlike denominators to common denominators and add.

1)

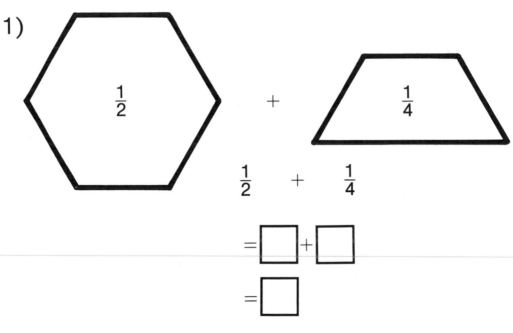

$\frac{1}{2}$ + $\frac{1}{4}$

= ☐ + ☐

= ☐

2)

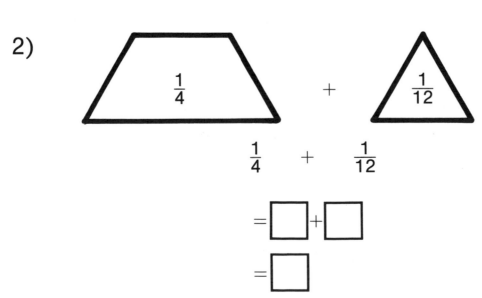

$\frac{1}{4}$ + $\frac{1}{12}$

= ☐ + ☐

= ☐

82

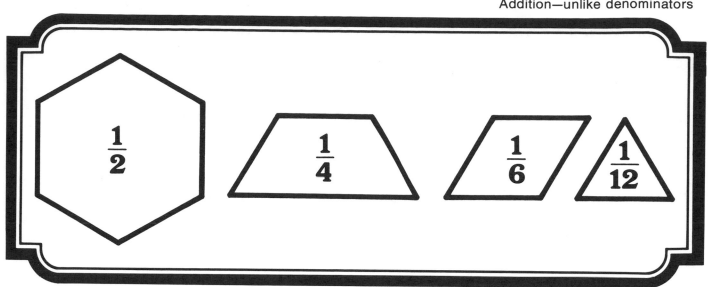

Change unlike denominators to common denominators and add.

Example: $\frac{1}{2}$ + $\frac{1}{6}$ = $\boxed{\frac{3}{6}}$ + $\boxed{\frac{1}{6}}$ = $\boxed{\frac{4}{6}}$

(yellow) (blue) (blue) (blue) (blue)

1) $\frac{1}{2}$ + $\frac{1}{4}$ = ☐ + ☐ = ☐

2) $\frac{1}{4}$ + $\frac{1}{6}$ = ☐ + ☐ = ☐

3) $\frac{1}{6}$ + $\frac{5}{12}$ = ☐ + ☐ = ☐

4) $\frac{3}{4}$ + $\frac{1}{12}$ = ☐ + ☐ = ☐

5) $\frac{1}{2}$ + $\frac{1}{12}$ = ☐ + ☐ = ☐

83

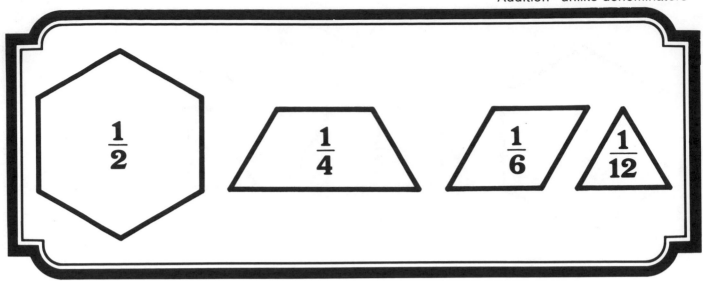

Change unlike denominators to common denominators and add.

Example: $\dfrac{1}{4}$ + $\dfrac{2}{6}$ = $\boxed{\dfrac{3}{12}}$ + $\boxed{\dfrac{4}{12}}$ = $\boxed{\dfrac{7}{12}}$
(red) (blue) (green) (green) (green)

1) $\dfrac{3}{4} + \dfrac{1}{6}$ = ☐ + ☐ = ☐

2) $\dfrac{1}{2} + \dfrac{5}{12}$ = ☐ + ☐ = ☐

3) $\dfrac{1}{12} + \dfrac{3}{4}$ = ☐ + ☐ = ☐

4) $\dfrac{5}{6} + \dfrac{1}{12}$ = ☐ + ☐ = ☐

5) $\dfrac{2}{6} + \dfrac{1}{4}$ = ☐ + ☐ = ☐

84

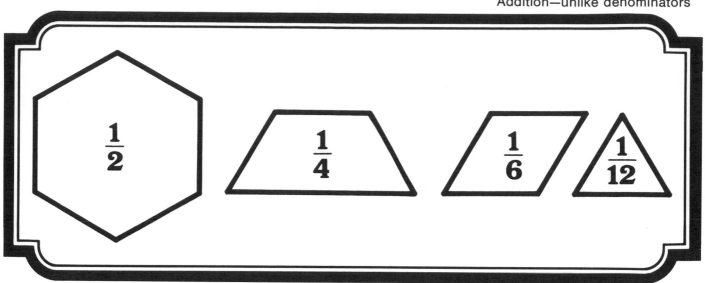

Change unlike denominators to common denominators and add.

Example: $\dfrac{1}{2}$ + $\dfrac{1}{4}$ + $\dfrac{1}{6}$ = $\boxed{\dfrac{6}{12}}$ + $\boxed{\dfrac{3}{12}}$ + $\boxed{\dfrac{2}{12}}$ = $\boxed{\dfrac{11}{12}}$
(yellow) (red) (blue) (green) (green) (green) (green)

1) $\dfrac{1}{2}$ + $\dfrac{1}{12}$ + $\dfrac{1}{4}$ = ☐ + ☐ + ☐ = ☐

2) $\dfrac{1}{4}$ + $\dfrac{1}{6}$ + $\dfrac{1}{12}$ = ☐ + ☐ + ☐ = ☐

3) $\dfrac{1}{2}$ + $\dfrac{1}{6}$ + $\dfrac{1}{12}$ = ☐ + ☐ + ☐ = ☐

4) $\dfrac{1}{4}$ + $\dfrac{2}{6}$ + $\dfrac{3}{12}$ = ☐ + ☐ + ☐ = ☐

5) $\dfrac{3}{4}$ + $\dfrac{1}{12}$ + $\dfrac{1}{6}$ = ☐ + ☐ + ☐ = ☐

85

Improper fractions

If the numerator (top number) of a fraction is larger than the denominator (bottom number) the fraction is called an *improper fraction*.

IMPROPER FRACTION: $\frac{3}{2}$ ◄── numerator / denominator

An improper fraction can also be written as a mixed number.

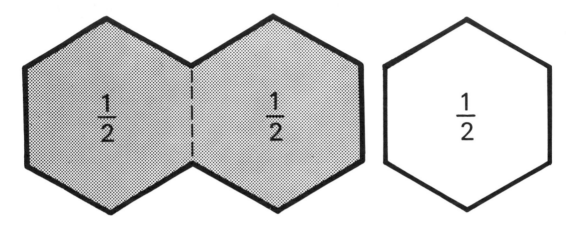

This is a model for $\frac{3}{2}$.

Three halves is the same as one and one-half.

$$\frac{3}{2} = 1\frac{1}{2}$$

Use pattern blocks to rename each improper fraction as a mixed number.

$\frac{7}{6} = \Box$ $\frac{11}{4} = \Box$ $\frac{16}{12} = \Box$

86

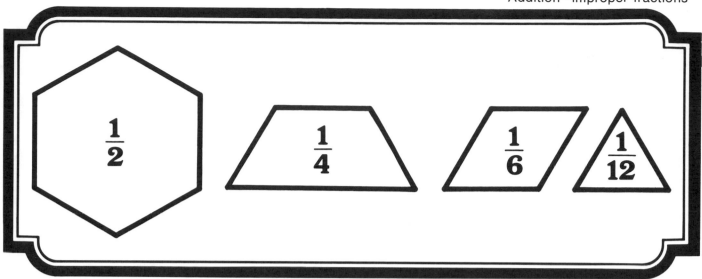

Add the fractions and write the sum as an improper fraction.

Examples: $\frac{2}{4} + \frac{3}{4} = \boxed{\frac{5}{4}}$

$\frac{4}{6} + \frac{5}{6} = \boxed{\frac{9}{6}}$

1) $\frac{7}{12} + \frac{10}{12} = \boxed{}$

2) $\frac{3}{4} + \frac{3}{4} = \boxed{}$

3) $\frac{1}{6} + \frac{5}{6} + \frac{5}{6} = \boxed{}$

4) $\frac{1}{2} + \frac{1}{2} + \frac{1}{2} = \boxed{}$

5) $\frac{5}{6} + \frac{3}{6} + \frac{2}{6} = \boxed{}$

6) $\frac{5}{12} + \frac{11}{12} = \boxed{}$

7) $\frac{3}{2} + \frac{5}{2} = \boxed{}$

8) $\frac{7}{6} + \frac{4}{6} = \boxed{}$

Write each answer above as a mixed number.

1) $\boxed{}$ 2) $\boxed{}$ 3) $\boxed{}$ 4) $\boxed{}$

5) $\boxed{}$ 6) $\boxed{}$ 7) $\boxed{}$ 8) $\boxed{}$

87

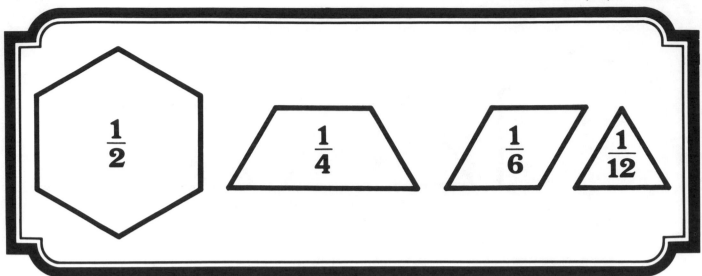

Solve the problems below.

Use pattern blocks to check.

1) $\frac{1}{2} + \frac{3}{4} =$ ☐ + ☐ = ☐

2) $\frac{3}{4} + \frac{3}{4} + \frac{3}{4} =$ ☐

3) $\frac{1}{2} + \frac{1}{4} + \frac{3}{6} =$ ☐ + ☐ + ☐ = ☐

4) $\frac{5}{12} + \frac{3}{4} =$ ☐ + ☐ = ☐

5) $\frac{5}{6} + \frac{3}{4} =$ ☐ + ☐ = ☐

6) $\frac{1}{2} + \frac{1}{4} + \frac{1}{6} + \frac{7}{12} =$ ☐ + ☐ + ☐ + ☐ = ☐

88

A fraction is in *simplest form* when it is written with the smallest denominator possible.

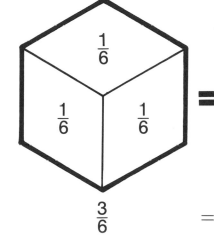 = 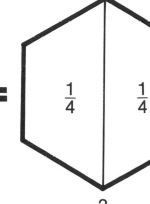 = 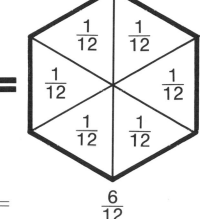 =

$$\frac{3}{6}$$ = $$\frac{2}{4}$$ = $$\frac{6}{12}$$ =

$$\frac{1}{2}$$

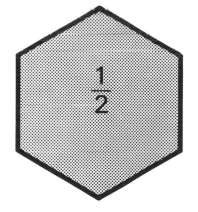

SIMPLEST FORM

One way to think of simplifying a fraction is to think of representing it with the fewest pattern blocks of one color.

Which is simpler, $\frac{1}{6}$ or $\frac{2}{12}$?

Which is simpler, $\frac{1}{2}$ or $\frac{2}{4}$?

Trace the blocks to show the answer.

89

Which is in simplest form? Circle the answer.

Use pattern blocks to check.

1) 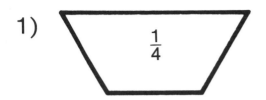 = 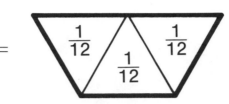 $\frac{1}{4}$ or $\frac{3}{12}$

2) 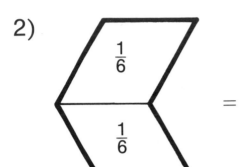 = 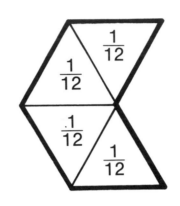 $\frac{3}{6}$ or $\frac{1}{2}$

3) $\frac{3}{4}$ or $\frac{9}{12}$?

4) $\frac{3}{12}$ or $\frac{1}{4}$?

5) $\frac{10}{12}$ or $\frac{5}{6}$?

6) $\frac{3}{2}$ or $\frac{6}{4}$?

7) $\frac{1}{6}$ or $\frac{2}{12}$?

8) $\frac{7}{6}$ or $\frac{14}{12}$?

9) $\frac{8}{6}$ or $\frac{4}{3}$ or $\frac{16}{12}$?

10) $\frac{10}{4}$ or $\frac{15}{6}$ or $\frac{5}{2}$?

90

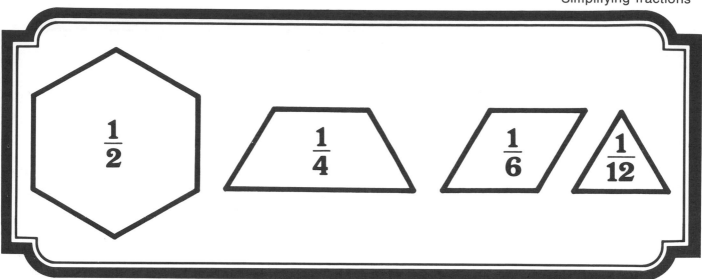

Build a model for each fraction below with pattern blocks.

Exchange for the fewest blocks of one color.

Write the fraction in *simplest form*.

Example: $\frac{3}{6}$ = $\boxed{\frac{1}{2}}$

(3 blue) = (1 yellow)

1) $\frac{2}{4}$ = ☐   6) $\frac{4}{4}$ = ☐

2) $\frac{2}{12}$ = ☐   7) $\frac{6}{4}$ = ☐

3) $\frac{3}{12}$ = ☐   8) $\frac{14}{12}$ = ☐

4) $\frac{3}{6}$ = ☐   9) $\frac{9}{6}$ = ☐

5) $\frac{9}{12}$ = ☐   10) $\frac{10}{12}$ = ☐

91

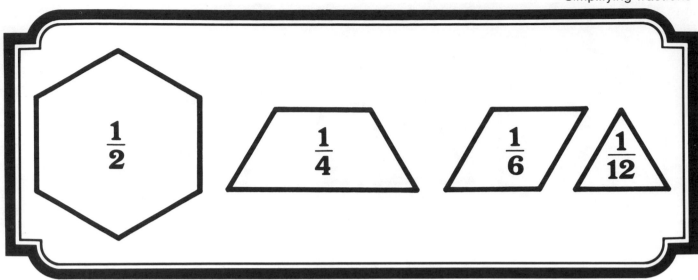

Add. Write the answers in *simplest form.*

Use pattern blocks to check.

1) $\frac{1}{4} + \frac{1}{4} =$ ☐

2) $\frac{1}{6} + \frac{2}{6} =$ ☐

3) $\frac{2}{12} + \frac{1}{12} =$ ☐

4) $\frac{2}{6} + \frac{6}{12} =$ ☐

5) $\frac{1}{2} + \frac{3}{12} =$ ☐

6) $\frac{1}{6} + \frac{1}{6} + \frac{1}{2} =$ ☐

7) $\frac{1}{4} + \frac{5}{6} + \frac{1}{12} =$ ☐

8) $\frac{3}{4} + \frac{3}{4} + \frac{1}{2} =$ ☐

9) $\frac{3}{4} + \frac{1}{6} + \frac{1}{12} =$ ☐

10) $\frac{4}{6} + \frac{7}{12} + \frac{1}{4} =$ ☐

92

Number of players: 2–4

Materials: Pattern block hexagons, trapezoids, blue parallelograms, triangles.
One die or a 1–6 spinner.

Rules:

1) Choose one player to be captain.

2) Give each player three hexagons.

3) Each player, in turn, rolls the die (or spins). The number rolled is the number of triangles that the player must pay to the captain. Each player must have the fewest number of blocks possible at all times.

Example: A player rolls four, trades one hexagon for six triangles, pays the captain four triangles, and trades the remaining two triangles for one blue parallelogram.

4) The winner is the first player to get rid of all of his blocks. Winner must *go out* exactly. To win, player must have exact amount rolled on die.

Subtract means "take away." Use pattern blocks to solve these subtraction problems.

Example: $\frac{12}{12} - \frac{1}{12}$ can be done like this:

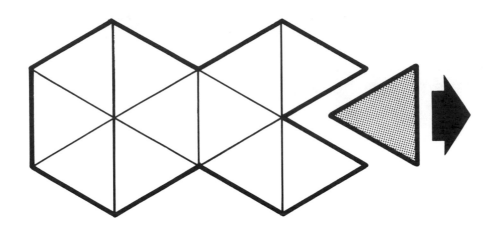

1) $\frac{12}{12} - \frac{11}{12} = \boxed{}$

2) $\frac{12}{12} - \frac{6}{12} = \boxed{}$

3) $\frac{12}{12} - \frac{5}{12} = \boxed{}$

4) $\frac{12}{12} - \frac{9}{12} = \boxed{}$

5) $\frac{2}{12} - \frac{1}{12} = \boxed{}$

6) $\frac{8}{12} - \frac{5}{12} = \boxed{}$

7) $\frac{10}{12} - \frac{7}{12} = \boxed{}$

8) $\frac{4}{12} - \frac{3}{12} = \boxed{}$

9) $\frac{20}{12} - \frac{10}{12} = \boxed{}$

10) $\frac{13}{12} - \frac{7}{12} = \boxed{}$

94

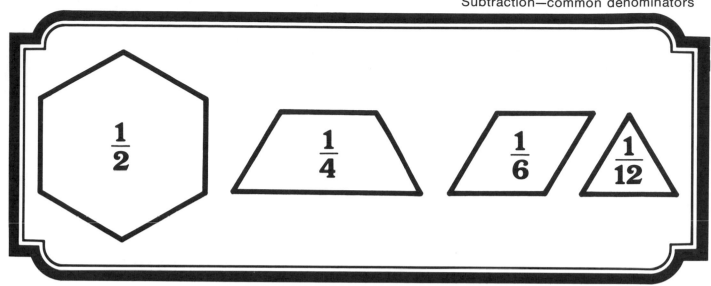

Subtract. Write the answer in simplest form. You may want to use pattern blocks to check.

1) $\frac{2}{6} - \frac{1}{6} = \square$

2) $\frac{5}{6} - \frac{2}{6} = \square$

3) $\frac{3}{4} - \frac{1}{4} = \square$

4) $\frac{7}{12} - \frac{6}{12} = \square$

5) $\frac{4}{6} - \frac{1}{6} = \square$

6) $\frac{10}{12} - \frac{7}{12} = \square$

7) $\frac{7}{6} - \frac{4}{6} = \square$

8) $\frac{5}{4} - \frac{3}{4} = \square$

9) $\frac{3}{2} - \frac{1}{2} = \square$

10) $\frac{7}{4} - \frac{6}{4} = \square$

11) $\frac{10}{6} - \frac{5}{6} = \square$

12) $\frac{83}{2} - \frac{72}{2} = \square$

95

Change unlike denominators to common denominators and subtract.

Use pattern blocks to check.

Example:

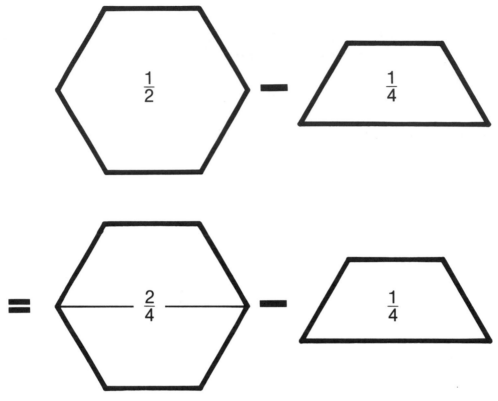

1) $\frac{1}{2} - \frac{1}{6} =$ ☐

2) $1 - \frac{7}{12} =$ ☐

3) $\frac{1}{6} - \frac{1}{12} =$ ☐

4) $\frac{3}{2} - \frac{3}{4} =$ ☐

96

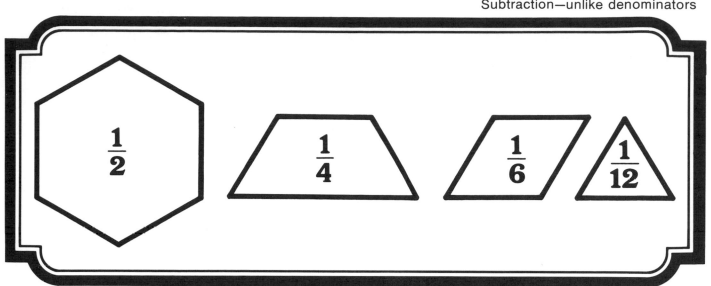

Change unlike denominators to common denominators and subtract.

Use pattern blocks to check.

Example: $\dfrac{1}{6}$ $-$ $\dfrac{1}{12}$ $=$ $\boxed{\dfrac{2}{12}}$ $-$ $\boxed{\dfrac{1}{12}}$ $=$ $\boxed{\dfrac{1}{12}}$

         (blue)   (green) (green)   (green)   (green)

1) $\dfrac{1}{2} - \dfrac{2}{6} = \Box - \Box = \Box$

2) $\dfrac{1}{2} - \dfrac{1}{12} = \Box - \Box = \Box$

3) $\dfrac{1}{4} - \dfrac{1}{6} = \Box - \Box = \Box$

4) $\dfrac{1}{4} - \dfrac{1}{12} = \Box - \Box = \Box$

5) $\dfrac{2}{6} - \dfrac{1}{12} = \Box - \Box = \Box$

97

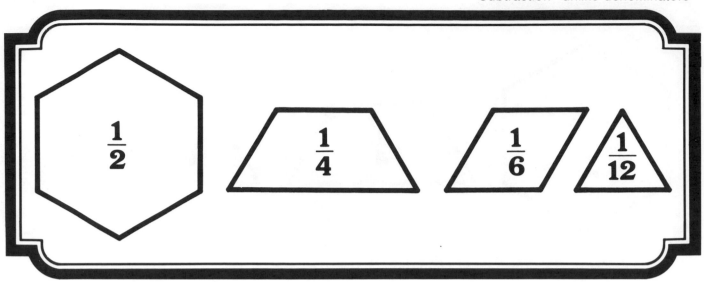

Change unlike denominators to common denominators and subtract.
Write the answer in simplest form.

1) $\frac{1}{2} - \frac{1}{4} =$ ☐ − ☐ = ☐

2) $\frac{3}{4} - \frac{1}{2} =$ ☐ − ☐ = ☐

3) $\frac{5}{6} - \frac{1}{2} =$ ☐ − ☐ = ☐

4) $\frac{5}{6} - \frac{1}{4} =$ ☐ − ☐ = ☐

5) $\frac{3}{4} - \frac{1}{6} =$ ☐ − ☐ = ☐

6) $\frac{7}{12} - \frac{1}{2} =$ ☐ − ☐ = ☐

7) $\frac{1}{2} - \frac{5}{12} =$ ☐ − ☐ = ☐

98

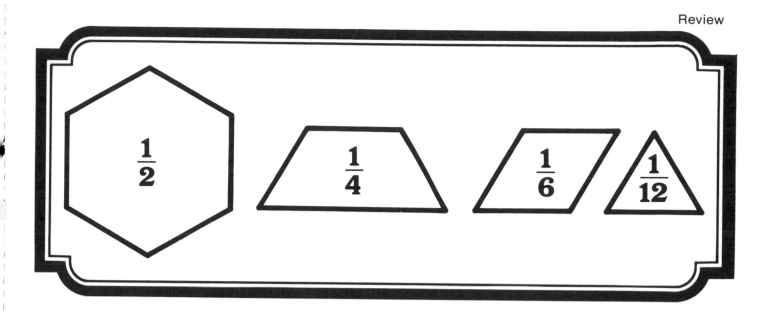

Solve the problems below.

Write the answer in simplest form.

1) $\frac{3}{4} + \frac{1}{6} =$ ☐

2) $\frac{3}{2} - \frac{1}{4} =$ ☐

3) $\frac{1}{4} + \frac{1}{2} + \frac{5}{6} =$ ☐

4) $\frac{3}{4} - \frac{7}{12} =$ ☐

5) $\frac{7}{6} + \frac{5}{4} =$ ☐

6) $\frac{3}{6} + \frac{3}{6} + \frac{1}{4} =$ ☐

7) $1 - \frac{1}{4} =$ ☐

8) $\frac{5}{4} - \frac{1}{2} =$ ☐

9) $\frac{7}{12} + \frac{5}{6} =$ ☐

10) $\frac{5}{2} - \frac{10}{6} =$ ☐

11) $2 - \frac{7}{6} =$ ☐

12) $1\frac{1}{2} - \frac{3}{4} =$ ☐

One-third is one of three equal parts that make a whole.

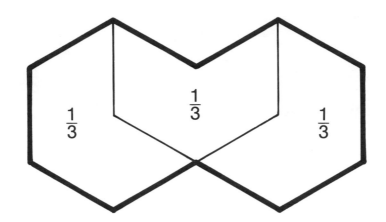

To make a block showing $\frac{1}{3}$, tape together two blue blocks.

(Since one blue block is $\frac{1}{6}$, $\frac{2}{6} = \frac{1}{3}$.)

Collect six blue blocks and make three-thirds.

Use the blocks to solve these problems. Write the answer in simplest form.

1) $\frac{1}{3} + \frac{1}{3} = \boxed{}$

2) $\frac{1}{3} + \frac{2}{3} = \boxed{}$

3) $\frac{3}{6} + \frac{1}{3} = \boxed{}$

4) $\frac{2}{3} - \frac{1}{6} = \boxed{}$

5) $\frac{1}{3} - \frac{2}{6} = \boxed{}$

6) $\frac{1}{2} + \frac{1}{3} = \boxed{}$

100

$\frac{1}{4}$ of 1 means one of four equal parts of one whole.

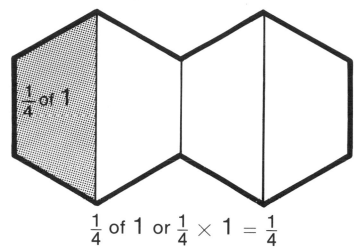

$\frac{1}{4}$ of 1 or $\frac{1}{4} \times 1 = \frac{1}{4}$

$\frac{1}{3}$ of 1 means one of _____ equal parts of one whole.

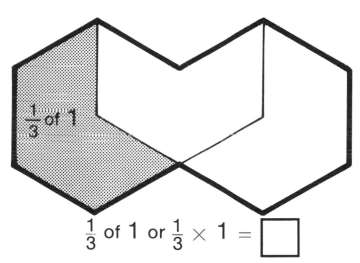

$\frac{1}{3}$ of 1 or $\frac{1}{3} \times 1 = \boxed{}$

$\frac{1}{6}$ of 1 means one of _____ equal parts of one whole.

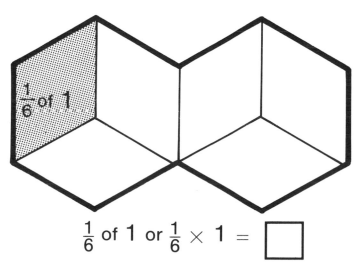

$\frac{1}{6}$ of 1 or $\frac{1}{6} \times 1 = \boxed{}$

101

$\frac{1}{2}$ of $\frac{1}{2}$ means one of two equal parts of one-half.

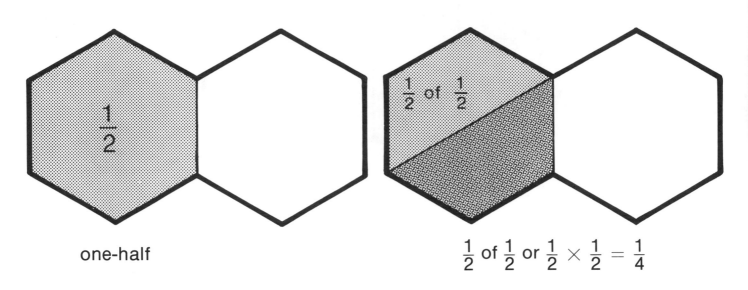

one-half

$\frac{1}{2}$ of $\frac{1}{2}$ or $\frac{1}{2} \times \frac{1}{2} = \frac{1}{4}$

$\frac{1}{3}$ of $\frac{1}{2}$ means one of three equal parts of one-half.

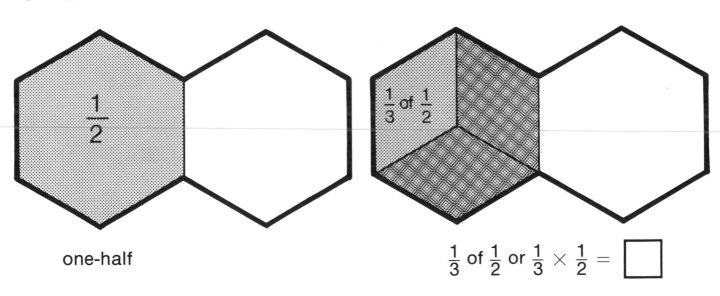

one-half

$\frac{1}{3}$ of $\frac{1}{2}$ or $\frac{1}{3} \times \frac{1}{2} = \boxed{}$

What is $\frac{1}{6}$ of $\frac{1}{2}$? Show it with pattern blocks.

$\frac{1}{6} \times \frac{1}{2} = \boxed{}$

102

$\frac{1}{2}$ of $\frac{1}{3}$ means one of two equal parts of one-third.

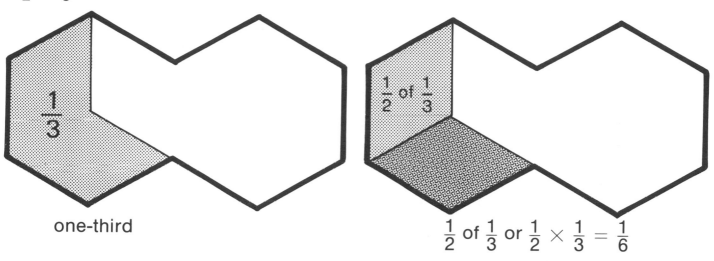

one-third

$\frac{1}{2}$ of $\frac{1}{3}$ or $\frac{1}{2} \times \frac{1}{3} = \frac{1}{6}$

Find $\frac{1}{3}$ of $\frac{1}{4}$. Use pattern blocks. Draw your answer.

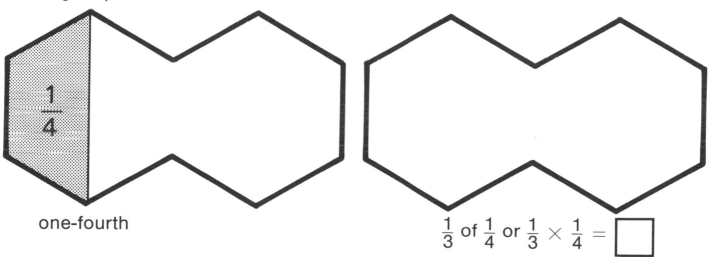

one-fourth

$\frac{1}{3}$ of $\frac{1}{4}$ or $\frac{1}{3} \times \frac{1}{4} = \boxed{}$

Find $\frac{1}{4}$ of $\frac{1}{3}$. Use pattern blocks. Draw your answer.

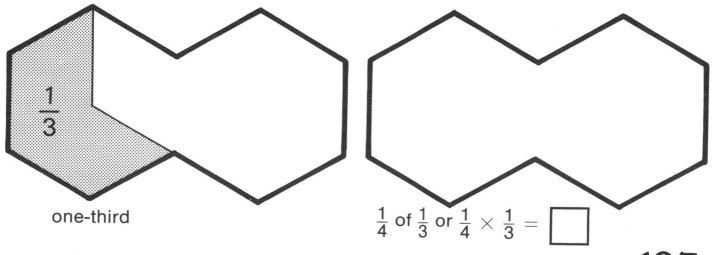

one-third

$\frac{1}{4}$ of $\frac{1}{3}$ or $\frac{1}{4} \times \frac{1}{3} = \boxed{}$

103

$\frac{2}{3} \times \frac{1}{4}$ means two of three equal parts of one-fourth.

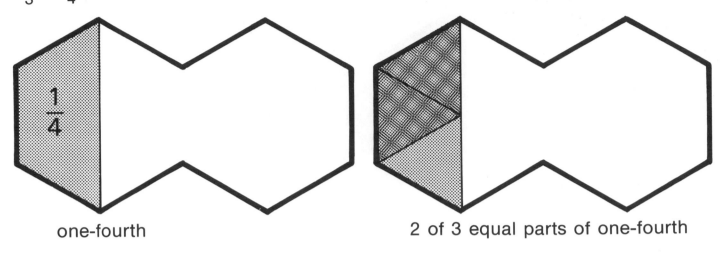

one-fourth   2 of 3 equal parts of one-fourth

$$\frac{2}{3} \times \frac{1}{4} = \frac{2}{12} \text{ or } \frac{1}{6}$$

$\frac{3}{4} \times \frac{1}{3}$ means _____ of _____ equal parts of one-third.

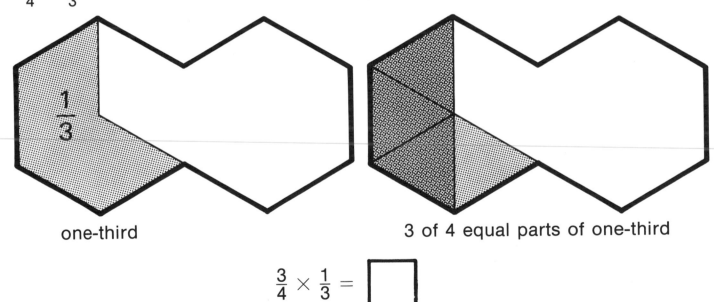

one-third   3 of 4 equal parts of one-third

$$\frac{3}{4} \times \frac{1}{3} = \boxed{}$$

Solve using pattern blocks.

1) $\frac{2}{3} \times \frac{1}{2} = \boxed{}$

$\left(2 \text{ of } 3 \text{ equal parts of } \frac{1}{2}\right)$

2) $\frac{5}{6} \times \frac{1}{2} = \boxed{}$

$\left(5 \text{ of } 6 \text{ equal parts of } \frac{1}{2}\right)$

104

Look for a pattern in the multiplication examples below.

$$\frac{1}{2} \times \frac{1}{6} = \frac{1}{12}$$

$$\frac{1}{6} \times \frac{1}{6} = \frac{1}{36}$$

$$\frac{1}{4} \times \frac{1}{3} = \frac{1}{12}$$

$$\frac{1}{4} \times \frac{1}{4} = \frac{1}{16}$$

Solve these problems using the pattern you found.

1) $\frac{1}{2} \times \frac{1}{2} =$ ☐

2) $\frac{1}{2} \times \frac{1}{3} =$ ☐

3) $\frac{1}{6} \times \frac{1}{2} =$ ☐

4) $\frac{1}{4} \times \frac{1}{6} =$ ☐

5) $\frac{1}{3} \times \frac{1}{6} =$ ☐

6) $\frac{1}{12} \times \frac{1}{12} =$ ☐

Look for a pattern in the multiplication examples below.

$\frac{2}{3} \times \frac{2}{3} = \frac{4}{9}$

$\frac{1}{2} \times \frac{2}{3} = \frac{2}{6}$ or $\frac{1}{3}$

$\frac{1}{6} \times \frac{5}{6} = \frac{5}{36}$

Solve these problems using the pattern you found.

1) $\frac{3}{4} \times \frac{3}{4} = \boxed{}$ 2) $\frac{3}{4} \times \frac{1}{6} = \boxed{}$

3) $\frac{3}{4} \times \frac{2}{3} = \boxed{}$ 4) $\frac{5}{6} \times \frac{3}{4} = \boxed{}$

5) $\frac{3}{2} \times \frac{2}{3} = \boxed{}$ 6) $\frac{5}{6} \times \frac{5}{6} = \boxed{}$

7) $\frac{7}{12} \times \frac{1}{2} = \boxed{}$ 8) $\frac{4}{6} \times \frac{2}{3} = \boxed{}$

9) $\frac{3}{4} \times \frac{1}{2} = \boxed{}$ 10) $\frac{7}{4} \times \frac{5}{6} = \boxed{}$

Describe the pattern you used.

_____

_____

106

One pattern used to multiply fractions is to multiply the numerators and multiply the denominators.

Example: $\frac{1}{2} \times \frac{2}{4} = \frac{1 \times 2}{2 \times 4} = \frac{2}{8} = \frac{1}{4}$

Use this pattern to solve the problems below. Write the answer in simplest form.

1) $\frac{1}{3} \times \frac{2}{6} = \square$

2) $\frac{1}{2} \times \frac{2}{3} = \square$

3) $\frac{4}{6} \times \frac{1}{4} = \square$

4) $\frac{3}{4} \times \frac{2}{6} = \square$

5) $\frac{1}{3} \times \frac{3}{12} = \square$

6) $\frac{3}{4} \times \frac{1}{2} = \square$

7) $\frac{1}{2} \times \frac{2}{6} = \square$

8) $\frac{2}{3} \times \frac{2}{3} = \square$

107

6 ÷ 2 can mean   In 6, there are how many groups of 2?

There are 3 groups of 2 in 6, so 6 ÷ 2 = 3.

12 ÷ 3 can mean   In 12, there are how many groups of 3?

Circle the groups of 3.

There are _____ groups of 3 in 12, so 12 ÷ 3 = ☐

Draw a picture which shows 10 ÷ 2.

In 10, there are how many groups of 2? _____

10 ÷ 2 = ☐

8 ÷ 4 can mean   In 8, there are how many groups of _____.

Draw a picture of 8 ÷ 4.

8 ÷ 4 = ☐

108

ONE WHOLE

1) $1 \div \frac{1}{2}$ means   In one whole, there are how many halves?

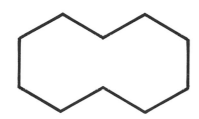 needs how many  to cover it? _____

$$1 \div \frac{1}{2} = \boxed{}$$

2) $\frac{1}{2} \div \frac{1}{6}$ means $+$ In one-half, there are how many sixths?

 needs how many to cover it? _____

$$\frac{1}{2} \div \frac{1}{6} = \boxed{}$$

3) $\frac{1}{3} \div \frac{1}{6}$ means   In one-third, there are how many sixths?

_____ needs how many _____ to cover it? _____

$$\frac{1}{3} \div \frac{1}{6} = \boxed{}$$

109

Write the division problem under the picture.

Use pattern blocks to find the answer.

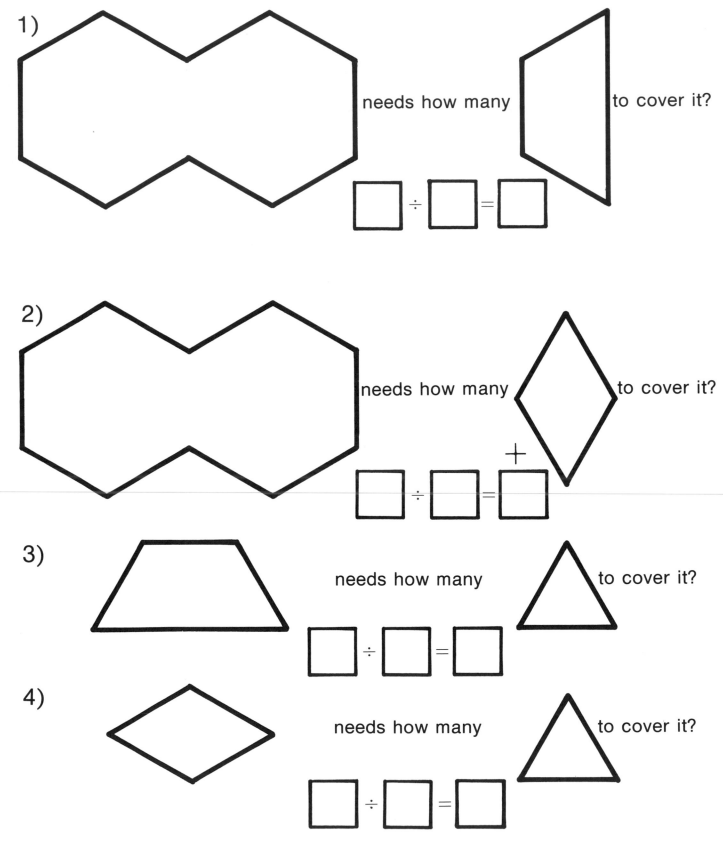

1) needs how many [trapezoid] to cover it?

☐ ÷ ☐ = ☐

2) needs how many [rhombus] to cover it?

☐ ÷ ☐ = ☐ +

3) needs how many [triangle] to cover it?

☐ ÷ ☐ = ☐

4) needs how many [triangle] to cover it?

☐ ÷ ☐ = ☐

110

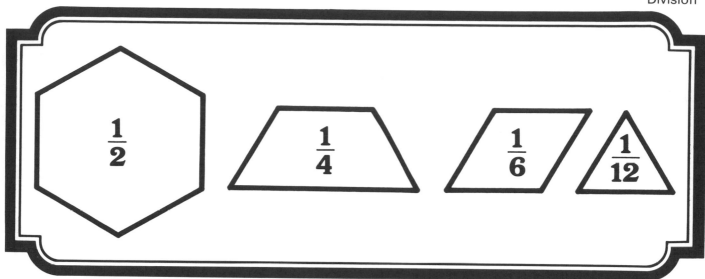

Use pattern blocks to complete the sentence and find the answer for each problem.

1) $\frac{1}{2} \div \frac{1}{6}$ means   In _____ there are how many _____?

$\frac{1}{2} \div \frac{1}{6} = \boxed{}$

2) $\frac{1}{3} \div \frac{1}{6}$ means   In _____ there are how many _____?

$\frac{1}{3} \div \frac{1}{6} = \boxed{}$

3) $\frac{1}{12} \div \frac{1}{12}$ means   In _____ there are how many _____?

$\frac{1}{12} \div \frac{1}{12} = \boxed{}$

4) $1 \div \frac{1}{3}$ means   In _____ there are how many _____?

$1 \div \frac{1}{3} = \boxed{}$

5) $1\frac{1}{3} \div \frac{2}{3}$ means   In _____ there are how many _____?

$1\frac{1}{3} \div \frac{2}{3} = \boxed{}$

111

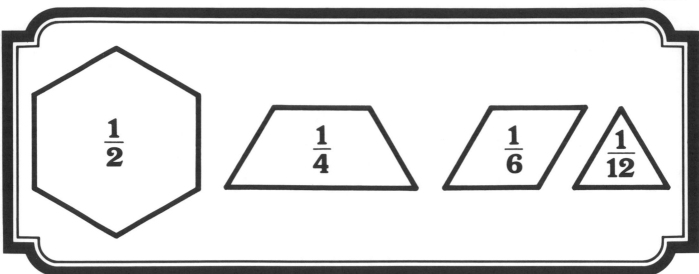

Use pattern blocks to solve these problems.

1) $\frac{1}{3} \div \frac{1}{12} =$ ☐

2) $\frac{1}{2} \div \frac{1}{4} =$ ☐

3) $1 \div \frac{1}{12} =$ ☐

4) $\frac{8}{3} \div \frac{1}{6} =$ ☐

5) $\frac{3}{4} \div \frac{1}{4} =$ ☐

6) $\frac{5}{6} \div \frac{5}{12} =$ ☐

7) $\frac{11}{12} \div \frac{11}{12} =$ ☐

8) $1\frac{1}{2} \div \frac{3}{4} =$ ☐

9) $\frac{3}{4} \div \frac{1}{12} =$ ☐

10) $\frac{14}{12} \div \frac{1}{6} =$ ☐

112

Sometimes division problems do not "come out evenly."

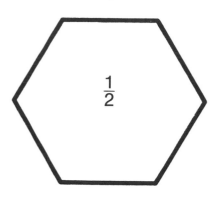

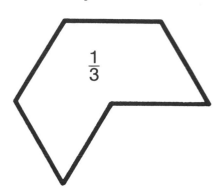

$\frac{1}{2} \div \frac{1}{3}$ means   In ⬡ there are how many ⬠ ?

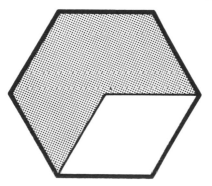

There is one whole ⬠ in ⬡ with some left over. The leftover is the *remainder*. The remainder is what part of ⬠ ?

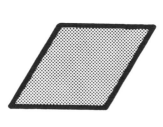   covers $\frac{1}{2}$ of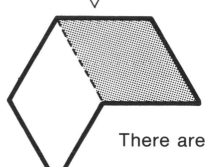

There are 1 and $\frac{1}{2}$ ⬠ in ⬡

so $\frac{1}{2} \div \frac{1}{3} = 1\frac{1}{2}$.

Use blocks to solve these problems.

1) $\frac{3}{4} \div \frac{1}{2} = $ ☐

2) $\frac{5}{6} \div \frac{1}{3} = $ ☐

113

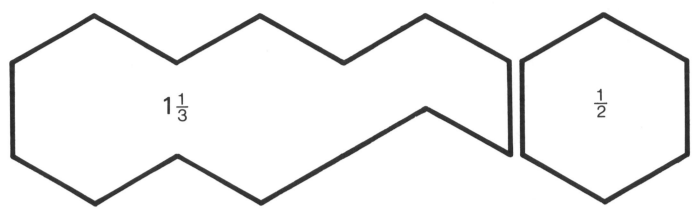

$1\frac{1}{3} \div \frac{1}{2}$ means   In ⬡⬡ , there are how many ⬡ ?

or, In $1\frac{1}{3}$ there are how many $\frac{1}{2}$'s?

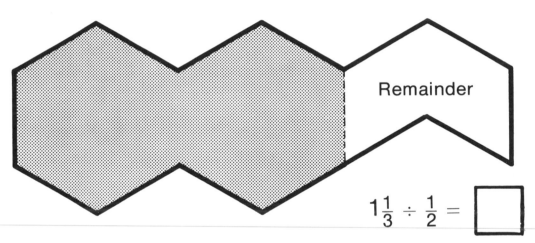

Remainder

$1\frac{1}{3} \div \frac{1}{2} = \square$

(Hint: Think, ⬠ is what fraction of ⬡ ?)

Use blocks to solve these problems:

a) $\frac{3}{4} \div \frac{1}{3} = \square$ b) $\frac{5}{6} \div \frac{1}{4} = \square$

114

These problems do not come out evenly.

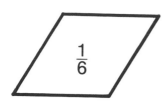 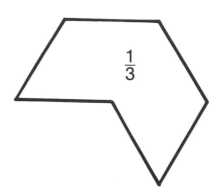

$\frac{1}{6} \div \frac{1}{3}$ means   In _____, there are how many _____?

will not fit in   even once. What fraction of will fit in ?

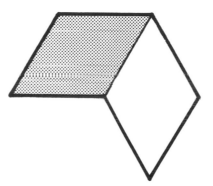

Only half of will fit in

So, $\frac{1}{6} \div \frac{1}{3} = \frac{1}{2}$

$\boxed{\frac{1}{6}}$ $\div$ $\boxed{\frac{1}{2}}$ means   In $\frac{1}{6}$ there are how many $\frac{1}{2}$'s?

Is there one whole ⬡ in ◇ ? _____

What fraction of ⬡ will fit in ◇ ? ☐

$\frac{1}{6} \div \frac{1}{2} =$ ☐

115

Divide using the pattern blocks. Don't forget the remainders.

1) $\frac{2}{3} \div \frac{1}{2} =$ ☐

2) $\frac{5}{6} \div \frac{1}{3} =$ ☐

3) $\frac{3}{4} \div \frac{1}{2} =$ ☐

4) $1 \div \frac{1}{3} =$ ☐

5) $1\frac{1}{2} \div 1 =$ ☐

116

Divide using pattern blocks.

1) $\frac{5}{6} \div \frac{1}{6} =$ □

2) $\frac{2}{3} \div \frac{1}{12} =$ □

3) $1\frac{1}{2} \div \frac{1}{2} =$ □

4) $\frac{1}{2} \div \frac{2}{3} =$ □

5) $\frac{2}{3} \div \frac{4}{6} =$ □

6) $\frac{3}{4} \div \frac{3}{4} =$ □

7) $1\frac{1}{2} \div \frac{1}{4} =$ □

8) $2 \div \frac{1}{3} =$ □

9) $1\frac{1}{2} \div \frac{1}{3} =$ □

10) $2\frac{2}{3} \div \frac{1}{6} =$ □

11) $3 \div \frac{5}{12} =$ □

12) $4 \div 1\frac{1}{2} =$ □